THREE LEVELS OF TIME

ALSO BY HAROLD HAYES

Smiling Through the Apocalypse: Esquire's *History of the Sixties*
The Last Place on Earth

THREE LEVELS OF TIME

Harold T. P. Hayes

E.P. Dutton—New York

Portions of this book originally appeared in other publications: chapter 7 in *GEO;* and chapter 12 in *The Atlantic Monthly*.

Published in the United States by
Elsevier-Dutton Publishing Co., Inc.,
2 Park Avenue, New York, N.Y. 10016

Library of Congress Cataloging in Publication Data
Hayes, Harold.
Three levels of time.

1. Ecology. 2. Human ecology. 3. Evolution.
4. Animals, Habits and behavior of. 5. Human
behavior. 6. Life—Origin. I. Title.
QH541.H28 1981 577 80-22963

ISBN: 0-525-21853-X

Published simultaneously in Canada by
Clarke, Irwin & Company
Limited, Toronto and Vancouver

Design by Nicola Mazzella

10 9 8 7 6 5 4 3 2 1

First Edition

For
Bill McIlwain

In passing from the past to the future, we pass from memory and reflection to observation and current practice and thence to anticipation and prediction. As usually conceived, this is a movement from the known to the unknown, from the probable to the possible, from the domain of necessity to the open realm of choice. But in fact these aspects of time and experience cannot be so neatly separated. Some part of the past is always becoming present in the future; and some part of the future is always present in the past. Instead of thinking of these three segments of time in serial order, we would do well to take the view of a mathematician like A. N. Whitehead and narrow the time band to a tenth of a second before and the tenth of a second after any present event. When one does this, one understands that the past, the present, and the future are in that living moment almost one; and, if our minds were only capable of holding these three elements together in consciousness over a wider span of time, we should deal with our problems in a more organic fashion, doing justice not merely to the succession of events but to their virtual coexistence through anticipation and memory.

—LEWIS MUMFORD,
Man's Role in Changing the Face of the Earth

CONTENTS

A NOTE BEFORE DEPARTING

The seventh grade was my first discovery of how hard the educational process could be. A bad time. Logarithms, declensions, syntax—levels of information I had never imagined. And there were too the sexual apprehensions of adolescence, a cruel coincidence of timing for which there had been little more warning. Mr. Redmond, who taught geometry, was coach of the scrub football team. All my teachers before had been gentle women. Mr. Redmond was smart and curt. If you asked a dumb question or gave the wrong answer he would sneer and throw chalk at the wastebasket. I became afraid to ask questions at all.

Mr. Phole, on the other hand, a fat man with a tiny ridiculous mustache, taught "band," an extracurricular course. You could rent a horn for two dollars a semester and he would teach you how to play it well enough to enter the beginners' band, and then you were on your own. My friend Jack Miller had rented a clarinet and I decided I would, too. "We don't have any more clarinets," Mr. Phole said. "Here. Take this trombone. You'll like it. You blow into the mouthpiece as though you're spitting out grape hulls. Like this"—and he pursed his lips—"phght . . . phght . . ."

In addition to band, Mr. Phole taught science. No one took him very seriously, certainly no one was afraid of him. Each year he was said to open his science class by taking a match from his pocket and making a great show of striking it. "Fire, boys," he would say. "Where did it come from?" Then he would blow it out. "And where did it go?"

I stayed with my trombone through college, lugged it about from one base to the next during military time, and stayed with music because of it ever after. I'm sorry I didn't take his science class, because I have never since been very comfortable with the subject. Fortunate indeed is the student who is led to learn through interest rather than fear or obligation. The Pholes of this world have been the forces of light for me, the Redmonds the forces of darkness.

I was thirty before I fully realized if I was to learn, in any case, I had to ask questions whether or not I would be encouraged to do so, and no matter how dumb I might seem by asking them. The broader and less specific the questions, of course, the greater the hazard in raising them. Moreover, when questions veer toward the full weight of human experience, the risk element increases proportionately to such an unbearable degree that most people, most of the time, don't ask them at all.

"Who am I?" and "How did I get here?" are perfectly reasonable questions in a limited context, but if you take away the limits, you've really got your hands full. At full throttle they are questions mainly for somebody else, for wise men and priests. The learning process has as much to do with recognizing the kind of answers you can handle as it does with asking brave questions. You have to be careful. In the interest of efficiency one tends to limit the scope of his questions to matters directly related to experience.

"I firmly believe that there is an essential continuity among physical, geological, and cultural processes, from the workings of the solar system to the workings of the Parent-Teachers Association in the Ann Arbor school system," writes biologist Marston Bates. Now that is pure Phole, honing in on something very big indeed. Inviting, mysterious, promising—potentially thrilling. Still, late-blooming question-askers circle the claim warily. What do we have to know to know more? And how hard is it to get past what we must know in order to find out *how* the solar system is connected to the Ann Arbor Parent-Teachers Association? Will it be worth the trouble?

Despite my own resolve at thirty and the fact that I had made journalism my business, big questions were no more tempting for me than for anyone else: I had learned to ask what I needed to know but my needs were predictable—until one morning a few years ago, on the terrace of the Norfolk Hotel in Nairobi, when a biologist forced me to deal with some facts which edged me dangerously beyond the world of personal experience.

The mammals of this earth (of which we are one species) are moving toward extinction. In East Africa, in the Serengeti National Park, a generously representative sampling of them is still to be seen within a wilderness setting: scores of lion and rhinoceros, hundreds of elephant, thousands of zebra and giraffe, half a million gazelle, a million or more wildebeest—this last a bovine, dull-witted creature which is the most important of all the animals to the Serengeti. All the others in one way or another are conditioned by its presence. And in this most natural of settings the survival of the wildebeest—just as it is for men living beyond the park boundaries outside the Serengeti—depends on grass, which the wildebeest eats. No matter how much or how little there may be of it, depending on the seasonal rainfalls, the wildebeest population adjusts to the grass supply in such fashion as to take enough but never too much. Each year the grass continues to grow as it has in the past for untold millions of years, years reaching back long before the arrival of man on earth. However, modern men who feed their cattle on the same sort of grass just outside the park have not in recent years been able to make similar adjustments. Their numbers grow, as do their cattle, but the cattle eat the grass to a point beyond its capacity to recover. *Their* land declines to desert, with starvation and death sure to follow.

How do the wildebeest know to manage their resources whereas man does not?

It is not a question of *knowing* on the wildebeest's part, the biologist said. It is their behavior in response to the laws of natural order.

I couldn't grasp it. But a man is more intelligent than a wildebeest in any event. Why couldn't these men, with their superior reason, see what they were doing to their own food resources and ultimately to themselves, and correct their errors?

A reasonable question, as it turned out, although hardly a very big one—or so I thought at the time. Rather, it seemed remote, a local problem of land usage in an exotic part of the world, a question more nettling than imposing but provocative enough nevertheless to cause me to write a book about it, and then to continue reading, however indiscriminately, what I could find on the subject.

In upstate New York where I live, at the Ulster County Community College library, I skimmed through the shelf on animal behavior until I came upon a book with a promising title—*Animal Nature and Human Nature,* by W. H. Thorpe, a British ornithologist. (You will meet him and hear more of his book farther on.) Thorpe was puzzled, too, but because he is an informed scientist, he was considerably less hesitant than I about pushing the implications of the contrast. "It is important that people should be shown both the characteristics in which the animal world approaches, and in some cases *greatly exceeds,* mankind in achievement," Thorpe writes (italics mine), and I wondered then, as I do now, at the shocking intimations of this quiet imperative: My God, was it remotely possible that animals *could* be smarter than men?

Of course it is unflattering to ourselves in the extreme for Thorpe or anyone else to suggest that we human animals may be anything less than the most exalted of creatures—until it starts to come clear how other animals make do. They conform to their environment whereas we tend to set the terms for ours. While we can fly into space or descend to the bottom of the seas, they are pretty much stuck where they are. On the other hand, they survive in collaboration with the environment, or have done so thus far, whereas we manage not only to spoil for ourselves the grasslands of sub-Sahel Africa, but to befoul most of the places we go, which now is everywhere. The air and the water turn bad. There is garbage under the ice in the Antarctic. Still, however profligate may be our ways, you can't say that animals are "smart" in the sense that we are. They are captives of environment precisely because they *aren't* as smart as we are. The man who saw connections between the cosmos and the people of Ann Arbor—the same Marston Bates—has more properly phrased the question I was struggling to form:

"Is man a part of nature," he asks, "or is he something different, apart from nature, a kind of organism with some control over his own destinies?"

I had underestimated the significance of the wildebeest paradox. That, too, was as large an issue as one might care to make of it. Modified, Bates's question could read: Is man a part of nature, like the wildebeest of the Serengeti? Or is he something that has come to exist outside of nature's constraints?

There is some urgency about this, too. The rain forests which encircle the earth are the property of sovereign powers (Brazil, Zaire, etc.) who have the right to do with them what they will. However, in many ways they represent a collective resource important to the survival of many species, including our own. Presently they are being cut down more quickly than they can grow back; some say they will be gone by century's end. So will the agricultural land in the U. S. if the conversion of it to other uses also continues at the present rate. Where then will our own food come from? Other concerns related to his question are more subtle but no less acute. Speaking of these things, another biologist said, "You should talk with Sir Otto Frankel. He's the man who first called attention to the fact that our very success in agriculture is destroying the gene base of the crops we use. To me this is the greatest threat after the atomic bomb." Innocent surrogate of the Ann Arbor P.T.A., I had never heard of the problem!

Here now had begun to emerge for me a very big question indeed. What followed from it? W. H. Thorpe believed you had to look *behind* the question to find the answer (if there is an answer). It would be useless to try to discover how far man "is unique and how far just an exceptional animal," he said, "without first examining the question of the origins of the lower animals and indeed of life itself."

So by now you can see what I had gradually and somewhat ingenuously let myself in for. If, that is, one should be inclined to accept Thorpe's condition—and I was. If my wildebeest/man question had any merit at all, one would have to go to the question preceding it—that is, where did wildebeest and man come from?—and thence to the question before that, that is, where did animals come from? And before animals, where did plants come from? Before plants, life? Before life, the world? With all the limits down, I was back to those two simple yet terrifying questions that confront us all: Who am I? How did I get here?

As it was the only way I knew, I thought I might approach these questions journalistically—that is, as a reporter interviewing the people who were most likely to provide whatever answers are currently available, such people necessarily being among the most informed people in the world. And this—the occasion for this book—I decided next to do, instinctively bracing for the sound of chalk hitting the wastebasket.

Long after my return, as I was reading what I could find to help me understand what I had been told, I came to, in F. H. T. Rhodes's wonderful little book *The Evolution of Life,* an account of the origin of the fern, one of the most primitive plants on earth. Drawings of some of the earlier species showed one of them standing straight-stemmed, like a hooded monk, its early fronds folded concentrically at top. The time then was late April, and there were such ferns in my garden, beginning to emerge from the cold, moist soil. They were four hundred million years old! In the earliest form we can be sure of, we ourselves have been here just over two million years.

Where you and I and ferns come from, and how we got here, and the consequences to all things by the circumstances of our presence form the basis for the story that follows, a story with a beginning and middle at least (and some say an end); with some beguiling characters; a variety of tensions and crises; an international setting; passions and fears and hopes and love; at least one real—unmistakably real—hero; and an uncertain outcome.

But there has been some trouble in putting it all together. The story moves not as I moved about the world (from Fort Collins west to Santa Barbara, Australia, Kenya, England, Russia and back to New York) but within the chronology of time, á la Thorpe and Bates. That is, from the beginning of time, as it is assumed to have begun, through the present, according to what is known about such things. (I caught a cold in Nairobi and lost it in Cambridge, and though references to it tend to scramble my scheme—Cambridge opens the book and Nairobi is halfway through—I left them in. They helped me keep my bearings; same cold, in any case.) The point was, I thought from the first, to try to keep it all as orderly as possible. What I had here were several levels of time, intersecting.

But as I sifted my research, arranging it along these lines, I discovered trouble of a more fundamental sort. The answers I had before me seemed only to be deepening the questions. I was being turned back to the place where I had begun—to the conundrums formed on the terrace of the Norfolk. Enter by the back door, so to speak, John Vihtelic, who had nothing to do with any of this. Or so at first I thought. I had met him under other circumstances in the small town of Whitehead, Michigan, where he had told me his story that now, two years later, began suddenly to force itself upon the story I was trying to tell. That this happened as it did I ascribed at the time to the mysteries of synchronicity and let it go at that. Whatever it was, however, I came to realize that in the human responses of Vihtelic to the unusual circumstance he had found himself in—at a different time and half a world away—there lay the components of an answer (or at least as close as I would ever come to an answer) to the questions that first set me in motion.

H.T.P.H.
Stone Ridge, New York

THREE LEVELS OF TIME

1

INTO THE HOLE

Mount Rainier, Washington

It was just after four now, and having turned in his boots and his day pack to retrieve his driver's license, which he had left as security deposit, Vihtelic headed his Mercury wagon south down the Cascade Range in the general direction of Mount Hood, a hundred and twenty miles away. He had hiked up Rainier to nine thousand feet, just above the cloud line but an hour or two short of the overnight station at Camp Muir, turning back while there was still enough light to make it down. He had seen what he wanted of Rainier, and since he still had a day and a half left of the weekend, he had decided abruptly he might as well drive over and see Hood, too, while he was at it. In two weeks he would be heading back east.

At Randle he stopped to gas up and check directions. If he followed the map he would have to take the expressway which lay to the west and added another hundred miles to the trip. Hood was right down the range. There ought to be some kind of connecting road between the two mountains, he thought, and if there was, he could make it the better part of the way by dark. Then he could spend the night in the wagon, as he had done last night, have most of tomorrow to hike Hood and still get back to Portland in time for a good night's sleep before reporting to work on Monday. That was important. Vihtelic's boss knew he had the company car for the weekend, a privilege he had no intention of abusing.

The attendant said there might be a road through but he wasn't sure. He directed Vihtelic to a tavern down the block where an old-timer told him Route 123 was what he wanted, an old logging road north of Trout Lake. It wasn't marked and it alternated between blacktop and crushed gravel, but it was a shortcut to Hood, running right down the range. He told Vihtelic he thought it was under construction but it was definitely open and he gave him instructions on how to find it.

Okay, Vihtelic thought, drifting on down the mountainside, he was in no hurry but he certainly didn't want to waste time; and when he came to 123, just a little tired but pleasantly so, he turned into it, settling back for an easy scenic cruise through the Cascade mountains in this second week of September with everything going his way. New job, and after the first three weeks, easily the best job he'd ever had: good salary, expense account, international air travel card, company car, good company—great company—and doing work he really knew something about. At St. Mary's, when the dialysis machine went out, he knew it so well he could make whatever part might be needed to fix it. He hadn't gone looking for the job; Drake Willock had come to him. From medical tech in Muskegon to field representative of a national concern in six weeks. Not bad.

Vihtelic slowed for gravel. The road was two-lane but very narrow and rougher than he had expected—mushy under the wheel, and at places part of the shoulder had washed away. Coming to a fork he looked for a route sign but there was none, so he chose the road with the heavier tracks. It was very beautiful along here, slipping in and out of the shadows of the forest. Conifers, nearly all conifers. Lots of hemlock. Whatever else, he wasn't sure. But very moist. Who was it who told him there was a rain forest in these mountains, or maybe in the Olympics farther west? Rain forests in Washington! Joseph, probably. He would know something like that.

And a new life. Next summer he would get married—not that his and Mary's life together would change all that much. They had known each other practically all their lives, anyway. In a town of three thousand everybody knew everybody else. Mary's father, Les, was principal of the high school. When Vihtelic stayed out his last semester to work at the marina, Les would just call home and if somebody said he was working, that was okay by Les. It wasn't so okay by

Vihtelic's father, however, who though now seven years dead, still puzzled and intrigued Vihtelic. His father was precise, orderly, and studious, and though he wasn't overbearing about it, that was pretty much what he expected from the rest of them. Vihtelic had never seen him without a shirt and tie except maybe in the mornings on his way to the bathroom. He was a loving man who would give his last dollar for you to go to a football game. But he was a hard man to get close to. He didn't have time for small things, or patience with those who did. You might tell him you were going out for football, as Vihtelic did once, and he would say, why don't you go out for nuclear physics?

His mother was different; she came from different circumstances. She had her own car at sixteen. Before she married his father, she had never fried an egg. *Her* father owned a sixteen-cylinder Packard and the family rode in the back while the chauffeur drove. She had an easier nature, probably because she'd had an easier time of it. When Vihtelic decided to take off for California his first year out of high school—something a lot of kids were doing at the time—his father thought it was a bad idea but his mother was all for it. . . .

The light was fading now. Vihtelic looked at his watch. Seven-forty. On his right the trees dropped down sharply into a dark ravine girdled by the gravel road, a tight hairpin turn hugging the embankment to the left except for a wood bridge which spanned the deepest part of the ravine. He heard the wheels rattling the boards below, thump-*thump,* thump-*thump,* and then he lost consciousness. The wagon edged toward the shoulder, moving along the curve until the right front wheel struck a rut, jolting Vihtelic into awareness. What's going on? he said to himself, what *is* this? The car moved on, wallowing about the shoulder and then, before he could regain control of it, lurching into another but larger gash a few yards on, hanging to the shelf of the road for a fraction of an instant before going on over to the right and then down, rolling side over side a hundred and seventy-five feet down the wall of the ravine. Falling and turning inside the car, expecting now to die, Vihtelic yelled inside his head: "*Jesus Christ! What's happening to me?*"

The Solar System

There may have been universes before this one; there may be universes other than this one. Either possibility is not much more than a guess. Nor is it possible to speculate with any degree of confidence what happened just the instant before the present universe began to take form—what was there before, or how long whatever was there had been there (if something *was* there), or what it was like. But that it was an explosion which set the present universe in motion is accepted now in most quarters as fact, and the time of occurrence is believed by those who try to measure such things to have been as recent as fifteen billion years ago, as far back as twenty.

An explosion of such proportions is difficult to imagine. It did not occur in a single "place"; it would have had to be everywhere, raising the temperature in the first hundredth of a second after detonation to 100 billion degrees Centigrade, and filling the universe with light. If the time should ever come when it is known whether or not the universe is finite, more will be known about the shape of the universe from the very first. If finite, the universe may have been 125 billion light years in circumference. If not, there would have been no size to it, any notion of measurement being out of the question.

However unimaginable this initial event may seem, there is commonplace evidence of it to be looked at in the most conventional of settings. The dotlike "snow" that often flickers across the television screen is the residue of the primeval explosion of creation, those stuttering speckles a form of radio static set in motion by the detonation, and now, billions of years later, still pulsing through space.

In the cooling down which followed the first instant of the beginning, gas began to form; and over time, galaxies emerged, aggregates of stars appearing as raindrops do from within the center of a thundercloud. There are now 100 billion or so galaxies, and within each galaxy, 100 billion or so stars. All are flying out from the ancient seminal explosion now lost to the billions of years past. Viewed from a great distance they take only certain forms—sometimes that of an ellipse or, often, of a spinning pinwheel. The earth's parent galaxy, the Milky Way, which is a hundred thousand light-years wide, speeds outward at a million miles an hour. Stars inside it come and go. The sun, the earth's parent star, is a second or third generation star and it has lived about half its lifetime.

Cambridge, England

The driver didn't know Cambridge although by his style of driving —turning purposively at intersections and holding to a constant speed —he was trying very hard not to let on that he didn't. We were running out of time, and reluctantly I suggested he stop and ask someone for directions. "Excuse me, sir," he said to a pedestrian, "can you tell me how to find the Cambridge Institute of Astrology?" Small comfort, I thought, as he moved on just as smoothly in a new direction, that my confusions were shared.

We come to the Institute, according to a small, tastefully lettered sign, at the outskirts of the university, tucked away behind hedgerows in the heart of an apple orchard, the only identifying clue from the road a stark white miniature observatory so unimposing as to reach barely beyond the tops of the apple trees. The weather is what the papers here describe as broken showers—mist alternating with soft rain. I asked the driver to wait and set off through dripping bushes to find Rees, or someone who could tell me where he might be. I was still plugged up from the cold, and though it was almost gone, I worried about getting my feet wet. Why risk bringing it on again?

There was no sign of life about the observatory—grass had grown too high all about it. Off to the side was a Victorian structure with oversized windows, the paint cracking on the doors, a gloomy, lifeless dwelling. Possibly a laboratory, I thought. I tried two of the doors—locked—and wandered on to a car park in the rear where a man was tinkering under the hood of his automobile. He pointed me back in the direction I had come, toward a modern building, an ambiguously rectangular brick box with picture windows. Martin Rees's office is a small room near the main entrance, very small—a cubicle really, less imposing for a stargazer than even the miniature observatory in the apple orchard. He is there waiting for me, and quite patiently as it turns out: in a few hours from now he is due to leave for Germany. But he urges me not to hurry. Would I care for coffee? An instant mix is produced, and he asks me to make myself comfortable.

To his back, and somewhat hesitantly, I refer to a letter I have sent ahead. In it, I said, I had wondered if he would tell me what is known about beginnings, about the progression of movements from the beginning of the universe through space leading to the beginning of the world. Where in this progression did the constituents of life come from? Of course I'm aware his time is limited—I begin to stumble—it is a lot to ask in so little time, I know . . .

There is no way to cushion the request. Rees smiles, passes the coffee, and acts as though I have asked nothing more exceptional than whether he thinks the sky will clear by evening.

"All you can suggest as you extrapolate back to earlier and earlier times—run the film backwards, as we say—is that the material in the universe gets squeezed to higher densities, becomes hotter and hotter—then it explodes. There is some speculation about what happened before that, of course—whether the explosion marked the end of an earlier universe, or whether ours is one of several universes coexisting, and so on. We don't know how material behaves under the circumstances of the instant of ignition. So there comes a point when one has to say that the physics as we know it becomes incomplete. Or inaccurate. Then even the boldest physicist will have to say that all bets are off. Some fundamental breakthrough is needed."

But the standard version of the beginning of the universe, after fifty years of research, has now become almost commonplace: a primordial fireball set things going. Somewhat playfully, George Gamow first described it as "The Big Bang." Although as a concept this is considered by most physicists to have been a near certainty, it is still not without some controversy. Briefly, Rees says, the background is this: Einstein's theory of relativity held that the universe is expanding; Edwin Hubble's observations added to this theory by demonstrating that the galaxies were rushing away from each other at enormous speeds. "That the expansion was initiated by some sort of explosion from a primordial dense state is the essence of the Big Bang theory. In the forties and fifties there was a rival steady-state theory which originated in this country with those who said, yes, the universe *is* expanding but it has always *looked* the same. They argued that new galaxies are being created all the time. As the old galaxies recede from each other, new galaxies appear, as it were, in the gaps so that things just don't become more sparse as time goes on. For example, if you took a snapshot of the universe in ten billion years it might look just the same as it does now. There would be different galaxies, you see, but the *number* of galaxies in a given area of space would be the same."

How to resolve such a disagreement? One way, Rees says, was to look as deeply as possible into the past. (When you look into the night sky, you are instantly looking into the past, of course. Though light travels very fast—186,000 miles a second—its speed is constant. So, when you look at something far away, you see it as it was when the light from it began its journey toward you. The distance light travels through space within a single year is six trillion miles. This is the arithmetic behind what is known as a "light-year." The light from a galaxy a billion light years away took a billion years to reach the earth.

Its distance would be 6,000,000,000,000,000,000,000 miles away. But you would *see* the galaxy as it was then, a billion years ago.) Looking into the past, Rees goes on, has become an increasingly sophisticated science. No matter how powerful their instruments, astronomers through the 1950s had trouble with the opacity of the earth's atmosphere. With the advent of space ships, the viewing device was shot beyond the atmosphere, with the consequence that there is at present such an abundance of new data that it threatens to overwhelm the researchers. In the twelve months between the recent launching of one space telescope and a later improved model, the information returned to earth gained by a factor of *thousands.* "That's like going in one decade from Galileo's telescope to the two-hundred-inch glass at Palomar," Rees says.

Anyway, getting back to the Big Bang versus the steady-state argument, "people like Allen Sandage and Martin Ryle studied distant objects in space to see if they were in any way systematically different from objects nearby. In a steady-state universe, all the galaxies should on average look the same. Sandage and Ryle claimed to find evidence that galaxies at large distances were different. The other, in a sense more important, development was the discovery by Arno Penzias and Robert Wilson of a background noise they were unable to account for in a radio receiver they had developed—a noise that was a curious hissing sound. Eventually they realized this was due to radiation coming from space—from all directions and from no obvious source. Not from the sun or from any known star or galaxy. The only tenable way to explain it was as a relic. A sort of cool echo of the primordial fireball." Then Rees told of the visible manifestation of this phenomenon in the static interference to be seen on an ordinary television set. *A cool echo!* Not a metaphor for the eye, but for the finger and the ear. Later I read that the imaginative Gamow, who earlier had suspected but failed to prove the existence of this microwave background radiation, had called it "the whisper of creation."

It has stopped raining. Like others engaged in his exotic specialty, Rees is an international commuter, having taught and researched at Harvard, Princeton, Cal Tech and the University of Sussex. At Cambridge he is Plumian Professor of Astronomy and Experimental Philosophy, a distinguished seat held previously by Fred Hoyle (steady-stater). If he gets away this afternoon as scheduled, he will lecture tomorrow in Bonn on the nuclei of galaxies. Tomorrow's audience may be disappointed, however, because of a slowdown of traffic controllers at Heathrow which promises inconvenience for me as well, since I am scheduled to leave London tomorrow. A travel inconvenience, so far as I can see, is all we have in common; we pause over our coffee in mutual commiseration, and I hold to the moment for a closer look at the man.

Late thirties, thin and angular, sharp-pointed features, restrained in manner but easy; voluble; accessible. I am genuinely grateful that he is willing to deal with me on a layman's level, and I tell him so.

"We are very fortunate as astronomers that people can see the stars overhead, which we work with, and that their curiosity becomes aroused by what we do," he responds politely, "so that there is some relationship between their interests and ours." Still idling, I mention the growing popularity of the notion there is life elsewhere in the universe, some astronomers suggesting the probable existence of a billion planets throughout our own galaxy where conditions for life would be as favorable as they are on earth. Rees shrugs. For him, it is not so attractive a possibility as one might expect. "Many people would say that if life *isn't* widespread in the universe, then the universe is a cold, inhospitable place with just one little speck of it in an inanimate cosmos. And this becomes ground for pessimism. I like the idea life might be a rare event and may even be unique to earth.

"If life is never destined to get beyond the earth, you can then say it doesn't matter cosmically if we're here or not. If there's life in the universe already, then again it doesn't matter. But if life exists only on the earth but does have the potentiality to spread through the universe, then it would be a cosmic disaster to snuff it out. You may say that life here looks very unprepossessing and we've goofed things up so much it wouldn't matter. But the answer to that would be: the first fish that crawled onto dry land might have looked rather unprepossessing. If you had clobbered it then, you'd have destroyed lots of interesting potentialities.

"Incidentally, this is the only sensible argument I can think of for supporting the sort of scheme of people like O'Neill"—Gerard K. O'Neill being the Princeton physicist who advocates self-supporting space colonies orbiting the earth—"which in a century could happen, in principle. Then the species as such would be invulnerable to any man-made disaster, such as biological or nuclear catastrophe wiping out life on earth. One can imagine life being destroyed on earth but not in a lot of independent colonies as well. So if one is concerned about the future of the species, then this line of reasoning starts to provide an argument. A pessimistic one perhaps, but it appeals to me. Because for the last twenty years we have been vulnerable to ecological or military disasters, and we seem to be in a position now where we can't make the world safe again."

There being not much more than a half hour left to cover some fifteen billion years, I wonder if we may get back to the fireball. Sure, he says, grinning.

Well then, after that, I say, still not without some discomfort at the inadequacy of my phrasing, what happened next?

A gas of hydrogen and helium spread thinly and evenly throughout the universe, and continued to do so for an indeterminable time, at the end of which the first stars emerged. A problem in setting distances and more exact times has to do with the limitations for astronomers in knowing for certain how far they can now see into the past. Perhaps eight billion light years, perhaps farther, Rees says. "We may be now looking back almost to the time when the first stars formed."

Whenever it was, he says, "stars began to form at the near edge of the void by condensing from diffuse gas clouds—clouds which had contracted because self-gravitation had pulled them together. You see, as the cloud contracts it heats up, and eventually it heats up enough to become luminous and starts to shine. Then it gets so hot the pressure building up inside it brings it into equilibrium as a star," a process which, he casually adds, still goes on. "Even quite near to us there's gas that hasn't yet turned into stars—dense regions of it like the so-called Orion nebula, where stars are still being born."

"This process can be seen?"

"Not with a single star, of course. But what you can do is photograph different stars in different stages of development, then put them together in a sequence. Imagine going into a forest without ever having seen a tree before. Unless you waited twenty years you wouldn't see a particular tree grow. But if the forest has been there a long time, you'll see seedlings and saplings and big trees. After only a few hours of observation you could then infer what the life cycle of a tree is. The relative number of them and their different sizes would tell you how fast they grow, for example. Cosmic evolution is very slow compared to the astronomer's lifetime, but the astronomer can still take advantage of the fact that there are millions of stars in the sky to photograph at different stages in their lives."

Stars die, like anything else, when they run out of fuel, which is hydrogen. It is from the process of nuclear fusion—turning hydrogen to helium at their cores—that stars derive their energy (a process some people are trying to duplicate on earth to supply our own needs for fuel). Hydrogen keeps a star shining, and when the hydrogen goes a star moves into crisis, seeking to force energy from what is left of it. Helium that has been processed from hydrogen is turned to carbon, and carbon is then turned to oxygen, nickel, and other elements, the last being iron, which is more nearly inert than any other in that its nucleus is the most tightly bound. When a star reaches that stage, no more energy is available to it.

The death of a star is a spectacular event. Its inner part implodes and may form a "black hole," which is a state of collapsed gravitation. The outer shell is blown off in a convulsion which flares to a brilliance billions of times greater than in its normal state.

It is the death of stars that brings us, Rees says, to the last of the questions I have raised—where does life begin?

"One of the main triumphs of astrophysics in the last twenty years has been to show that the relative amounts of different substances produced in the course of a star's evolution agree fairly well with the relative amounts that are observed on earth," he says. "For example, you find there is about three times as much oxygen as carbon in a sample of the earth, and you can show that's what you might expect in the course of a star's evolution."

Rees taps now at his desk, a pitifully tiny reference to the prospect of the sequence he is describing. "All the carbon in this wood, in our bodies, and in everything else on earth was synthesized from primordial hydrogen and helium in the stars which formed. Then the solar system condensed from the supernova that went off earlier in the galactic past. So the history of individual carbon atoms in any bit of living matter will take you back to the early history of the galaxy when that particular atom was forged from primordial hydrogen and helium in the incredible temperatures and densities inside a big star."

People—you, I, Martin Rees—and the world we inhabit became possible only after the first stars began to explode, in the thermonuclear reaction attending their exhaustion. If what is sought is the earliest occasion for the eventual existence of life—for the origin of the *constituents* of life—it is here, Rees says, in some unknown star's death that the true beginning is to be found.

2

COMING TO

Trout Lake, Washington

The overhead light was on but it was somehow *under* John Vihtelic. The better part of his weight rested against a crumpled bulge in the roof. He was facing toward the rear, half draped over the passenger's side of the wagon, pressing down on the bulge. He waited for pain. When he felt none he put his hands on the top of his head, then quickly ran them down his body to see if he was bleeding. He knew he would die from bleeding faster than anything else, faster than loss of air—anything. He found some scratches but he wasn't bleeding. He had better get out and away from the car as fast as possible. He could move the trunk of his body, both his arms and his right leg, but as he tried to push toward the window to his left he discovered that his left foot was pinned behind him somehow, and when he tried to pull it free the pain hit. It was unbearable. He couldn't move his left foot. He was pinned. What it was that held him he didn't know but he was pinned inside the wreckage of this car at the bottom of a ravine, in darkness, and however he was pinned the pain was excruciating. He strained against his left foot and the pain increased. Reaching behind him he pulled and tore at his left sneaker and at the same time he screamed, "Please help me! Somebody help me!" Within seconds he had torn the sneaker to pieces. He made himself stop and think. The car could catch on fire. He reached across to turn off the ignition. What he had to do first of all was get control of himself. He was probably in shock. He reached over his head to feel for the lever to the front seat and by pulling down on it he managed to push the seat back to give his head more room. The pale light beneath him showed all the familiar references reversed, turned upside down. Where he had been sitting—the driver's seat—was mashed flat, the roof crushed down to the steering wheel. If he'd had his seat belt fastened he would be dead.

The chassis was tilted slightly to the left, resting on something underneath. He could see nothing outside the car but suddenly, as though his mind now was willing to receive it, he heard the sound of a rushing creek or stream, so loud it surprised him that he had only now become aware of it. It was close by but how close he had no idea. He looked at his watch. Eight o'clock. Someone would be along soon and get him out. His right leg was doubled beneath him, his left stretched out behind him. He was held as though in the ready position of a sprinter awaiting the starting gun, except that the toes of his left foot, instead of bending under it, were stretched flat by whatever held them, pulling against the tendons of his leg. Pushing himself back to a half-kneeling position, as high as he could raise his body under the seat above him, and twisting to look over his shoulder, he could see how he was caught. A tree root about six inches in diameter and with a fork remnant at its base had penetrated the windshield. At a point between the dashboard and the root, the weight of the car was held by his foot. The car tilted because it had come to rest on Vihtelic's left foot and on the stump beneath it.

Vihtelic told himself to calm down. There was no point in getting excited. His foot was probably broken but if that was the worst of it, he could deal with that. It wouldn't be very long now before he was missed or someone else came along. He wasn't the only one to know about 123. Tomorrow was Sunday. A lot of people would be out. What he had to do was calm down, make himself as comfortable as possible, and wait. What on earth would they think at Drake Willock? Three weeks on the job and he had smashed up the company car. Nice going.

What he really had to do was *calm down* and make himself comfortable, start assessing his situation. He made himself think back to what he had started out with. Two sleeping bags, an air mattress, a suitcase with some extra clothes—he could use those now—his toiletry kit, a cooler of Cokes and beer, a coffee pot, two metal hot-dog forks, a paring knife, a camera, his tennis racket. As for food—a sack of apples, a package of rolls, some cheese crackers, a loaf of wheat bread. He wasn't hungry but he could use the clothes. The stuff was scattered out all to hell and gone.

He found an apple and ate it. He found an empty Coke can . . .

Why had he blacked out on the bridge? He remembered he was just coming off—he'd heard the boards rattling—and then he went out—not long, maybe two seconds; the jolt of the first washout had brought him to. Why? He was in good shape, he and Mike had jogged four miles a night to get ready for the hike, before Mike decided to go into Seattle instead, he—Vihtelic stopped on the connection. Anoxia, oxygen deprivation. In the past twenty-four hours he had moved from sea level to nine thousand feet up a mountain and then down again to—he guessed he was at about two thousand feet now, and he had just conked out, just long enough to drift to the edge. Then the washout had got him.

Glancing down, he noticed an edge of the package of rolls. It was just under the car frame, but the cellophane was torn and the rolls were wet. He tossed them aside. He couldn't find the paring knife or his toiletry kit. This last particularly annoyed him; he knew he'd better try to keep neat. The suitcase was within reach but his clothes were scattered. He found a turtleneck shirt, two T-shirts and a windbreaker, and he put them all on. He guessed it was about forty degrees, very damp. He could reach one of the sleeping bags and he wrapped it around himself as best he could. Important to stay warm.

Every so often he pulled at his foot, and each time he did the pain grew worse. At nine o'clock the overhead light burned out. He was glad that his watch had luminous hands. At ten o'clock, he thought: I've screwed myself out of my job. Nice job while it lasted. But that was the worst of it. Any minute now somebody would be coming up to the car saying, we found you. It would just be a matter of hours, that was all there was to it. There was no point in thinking about anything else.

The Azoic

Ten billion years ago, within the Milky Way galaxy, a large, cold, interstellar cloud of gas began to collapse—the initial event in the formation of the sun. First the cloud broke down into smaller clouds, and then one of these clouds, spinning like a ballet dancer tucking in her arms, condensed into the sun, a young star with so intense a concentration of energy at its center as to power a thermonuclear reaction and thus to provide, from then until now, and as far again into the future, a vast source of heat and light. Trailing in the orbit of the new sun was a disk of left-over debris—of ice, dust, and gas—and from this disk, four and a half billion years ago, nine planets took form, including the earth.

For the first ten million years of its existence, the earth was molten hot. (At its center, it still is; a temperature of 7,000 degrees Fahrenheit.) After ninety million years the earth had cooled down, and after a billion years, its surface skin had formed: an irregular contour of mountain ranges and basins rising and falling in response to the thrust of pressures under the crust. From volcanic gases escaping the heat in the earth's central furnace, like steam from a boiling pot, the atmosphere and the oceans began to form.

The earth is composed of silicate rocks and metallic iron, and it is now, as then, in constant motion, spinning on its own axis as it circles the sun, and moving with the sun around the center of the galaxy. From heat inside it and out, there is motion on it and within it. Volcanoes erupt, mountains and even continents move, driven by the convection currents produced by radioactive heat at its center. Heat from the sun keeps the oceans and the atmosphere moving; and the pull of the moon and the sun raises the tides twice a day.

Within the first billion years, life formed. The evidence of it has been found in rock fossils, impressions of microorganisms in deposits of chert (a form of sedimentary rock) in Ontario, South Africa, Greenland, and elsewhere, the oldest of these fossils having been dated at 3.8 billion years. The prevailing belief is that life emerged from nonliving matter in this way:

In the birth of the sun, as in the birth of other second-generation stars, the basic ingredients of matter collected: hydrogen, carbon, oxygen, phosphorus and other elements. In the birth of the earth from the debris left over from the birth of the sun, these elements were present in various proportions which, under the ultraviolet light of the sun or from lightning in the skies, produced a variety of combinations—in a phrase, organic molecules.

By the time a billion years had passed, two large molecules of unusual character and function had emerged. These were the molecules of nucleic acid and protein. The separate parts of these two molecules are chemically simple, but the way they work is complex. They re-create themselves.

In effect, the nucleic acid molecule orders the construction of the protein molecule, which then assists the nucleic acid molecule in copying itself. The most perplexing question raised by this, is: how? How could the two molecules produce others of their kind without there having been an earlier pair of them to perform the same task? Thus far the answer is unknown. Nevertheless, and however it happened, this is believed to be the point at which life on earth began.

Moscow

Nine-fifty-four A.M., chilly and raw; Leninsky Prospect—as wide as one remembers from the films of May Day parades, wider even than the Champs Elysées. The traffic now is falling off, only an occasional truck passing, black diesel waste trailing after, here as everywhere else. Except for the building behind me, a rambling stucco complex known as the A. N. Bakh Institute, I have no idea what part of the city's activities may be located hereabouts. Government? And if not government, what else? A pedestrian hurries past, brisk and as distracted as a New Yorker but toward a destination for me unimaginable. It was a mistake to send the Intourist hire car back before checking inside; now I have three hours to kill, here on the sidewalk, doing nothing. I am even unable to pretend to myself I am an informed observer. I am a happenstance visitor to Moscow, for the first time, without so much as a travel folder. Referenceless, I have no way to establish bearings. The streets don't look much different from ours, and the people by outward appearance aren't much different either. The weather is gray and damp as it has been everywhere else. I might as well be standing in a shopping mall in Akron. A drive off the main boulevard leads to a parking area beside the institute, and occasionally, I note, a car pulls in, close enough that I can get a glimpse of its occupants. I strain to see if there is an old man inside—he would have to be *very* old—on the off chance that he may be arriving before he is expected. After three cars over twenty minutes, I give it up and hitch myself onto the stone wall facing the sidewalk. What the hell, I will review my notes.

What is known about how life on earth began?

Evidently you must start with a catch-all explanation used by scientists from the earliest of times through the nineteenth century—

"spontaneous generation," a phrase that more or less defines itself. Out of a variety of circumstances, things came together in a certain way, and life just happened. A strong indication that this was so was the presence of the worm inside the apple. How else could it have got there? Through the process of spontaneous generation, Aristotle believed, the dew produced fireflies. The barnacle goose came from barnacles, birds from fruit trees, and lambs from melons. Filth, ooze, and mud were the most likely places you might look to see life begin. Descartes believed in spontaneous generation, and so did Newton. One Brussels physician, a man named Van Helmont, set forward a recipe for the spontaneous generation of mice: "If a dirty undergarment is squeezed into the mouth of a vessel containing wheat, within a few days (say twenty-one), a ferment drained from the garments and transformed by the smell of the grain, encrusts the wheat itself with its own skin and turns it into mice." Life arose from spontaneous generation, many earnest men have believed, as fire from spontaneous combustion.

Still, this was far from an elegant explanation, and, for some, hardly a satisfactory one. But the only evident alternative—that life is eternal —was equally hard to take. Expressing a typical frustration, Thomas Huxley a hundred years ago suggested that if it should be possible to reduce animate matter to its inanimate components, it ought to be possible to put it all together again. But—how? In Huxley's time the search for a more acceptable explanation gained in momentum with increasing refinements in the analytic methods of "reductionism," the study of separate parts to explain the workings of the whole. Nobody believed by then that mice were born out of old clothes, but the microscope *did* show tiny creatures emerging new and whole, seemingly, from hair, or a water drop, or the most unlikely of objects. Spontaneous generation persisted as the only possible explanation until Louis Pasteur himself set about disproving it. Through a series of laboratory experiments, Pasteur demonstrated the presence of microorganisms in the air. Life existed in forms so minute that one could not previously have suspected it was there—existed even in the air we breathe! Living things did not just suddenly appear; they came from other living things of the same kind—more precisely, of the same species. "Never will the doctrine of spontaneous generation arise from [my] mortal blow," Pasteur declared emphatically. This was in 1870. Fifty-three years later, in 1923, a monograph appeared in the Soviet Union entitled *The Origin of Life on Earth,* which in effect reopened the question by pushing it back some several billion years. Its author argued that there is no fundamental difference between living and nonliving matter. His name was Aleksandr Ivanovich Oparin, and he is the man I am killing time waiting to see.

From the Xerox copy I had made of an entry in a reference biography, he looks a cross between Walter Slezak and Nikolai Lenin, more serious than the former, less obsessed than the latter: in any case, an unlikely Frankenstein. There was no date on the picture; it could have been taken in 1938, when his work was first published in English. How ancient would he be today? Born in 1894, the reference says: eighty-four. He is still actively engaged in research, still the director of the A. N. Bakh Institute of Biochemistry behind me. Given his importance to the very first question of earth life, it's odd he is not better known in the West.

"At dinner he has a bottle of brandy on one side of his plate and a bottle of vodka on the other," an American associate once observed. "By the last course he has finished both." A man evidently of great vitality. In the Russian scientific community, which over the years has been bumpy going for its members, he is seen as a senior survivor. Once at a dinner of scientists in America, someone asked his position toward the dissident Sakharov. Madame Oparin, who teaches English at a university here and who serves as his translator abroad, conferred with him briefly and then responded, "Of course there are many Sakharovs in Moscow."

In his monograph of 1923, Oparin reasoned that the atmosphere of the primitive earth—the earth of four billion years ago—differed from what it is today. Hydrogen, carbon, nitrogen, and oxygen are the main components in the tissues of all living things. While the first three of these were present on the primitive earth, there was no oxygen—that is, "free oxygen"—as there is now (a key factor in his reasoning). That would come only after the arrival of plants, whose own life processes would release it into the atmosphere. Over hundreds of millions of years, molecular combinations of these several elements would fuse in reaction to ultraviolet radiation from the unfiltered sun, or heat from lightning, and in the absence of competition from other living organisms, life would begin.

Some scientists still find this theory hard to accept. Francis Crick, who with James Watson explained the self-replicating structure of DNA, believes it more likely that the earliest forms of life were brought to earth by other beings from outer space. One problem in even conceiving of such a sequence is the difficulty in imagining a time span of eons. But if you take the model of a lottery, Richard Dawkins suggests by way of illustrating the stretch, you can begin to see how *chance* works with time. In any man's lifetime, things which are improbable seem impossible, and that is why you never win the jackpot. But if it were possible to buy a chance every week for a billion years, you would probably win several times. Over a billion years, even the improbable becomes possible. For Dawkins and the majority of biologists, Oparin's ideas have become the standard version. They fit more comfortably with a natural order of things.

"Oparin's book wasn't translated into English until 1938," the same American associate had said, "and while it caused great excitement among graduate students in biochemistry, the war was coming and nothing could be done about it over here. Then in 1949, the cyclotron at Berkeley became available to the biologist Melvin Calvin. The first thing Calvin did was to run an experiment related to the primitive atmosphere. He took carbon dioxide, water, and hydrogen and exposed this mix to alpha particles from the cyclotron, and he got some simple compounds—formic acid, formaldehyde, very simple ones. He published his results in *Science,* in 1953.

"At that time Harold Urey was lecturing on the primitive atmosphere at Yale. Harold had worked out all the equilibrium constants to show that the primitive atmosphere had to be reducing—that is, nonoxidizing—so carbon dioxide had no place in it. And this of course was the argument Oparin had put forth. The carbon compounds had to be reduced. You see, in all the organic compounds you have in living systems, the carbon is joined to hydrogen, but if you had to start with something joined to *oxygen,* then to go back to hydrogen would be difficult. So Oparin suggested that the components had to be joined with hydrogen. It was a very simple idea based on biochemistry; and Urey had come to the same conclusion, based also on his consideration of astrophysical factors.

"So when Urey saw Calvin's paper, Urey was mad—mad because Calvin referred to Oparin's reducing atmosphere but used a *non*-reducing atmosphere. Urey published a paper in the proceedings of the National Academy attacking and correcting Calvin. And then he went back to Chicago—he had been at Yale for a year—and looked for somebody to do an experiment."

There he enlisted a graduate student, Stanley Miller, who constructed a laboratory apparatus which enabled him to simulate the early earth atmosphere described by Oparin. For a week Miller subjected it to electrical sparks—simulated lightning. By the end of the week he found in the bottom of his test tube some of the complex organic molecules needed for the organization of life.

"Lightning in the primordial soup" is the shorthand metaphor for this classic experiment, as familiar an expression to biologists now as Gamow's Big Bang is to physicists. (Among the exhibits at the Smithsonian Institution in Washington not so long ago was a repeating videotape of Julia Child, the culinary expert, describing and preparing Oparin's recipe for primordial soup.) Stanley Miller's experiment has since led to even more sophisticated variations which have yielded amino acids found in protein molecules, and sugars and phosphates found in the nucleic acid molecule.

As a consequence of these several developments, Oparin still labors at proving the ideas he conceived half a century ago, although on Mondays—at least by my experience acquired this Monday on Leninsky Prospect—he doesn't start work until noon.

He is seated behind a large desk in a comfortable office within this large building, which seems to be deteriorating. The wooden stairs leading here sagged, and the long hallways were dark; there were holes in the plaster ceilings. But the office turns out to be warm and well appointed, a privileged place. Oparin is smiling politely. I couldn't have spotted him from the street, though his appearance is certainly distinctive: a vast man, he is dressed rakishly in a salt-and-pepper tweed, with a large floppy bow tie and lightly rimmed glasses. True to the ancient picture I carry, he has a goatee, like Lenin's. There is a slight tremor to his hand but otherwise he looks to be in late middle age.

Seated between him and the deputy director of the institute (who will remain silent throughout the interview) is his translator, a young man in his twenties, slender and pale, who begins by apologizing (with, as it turns out, just cause) for his English. He welcomes me on behalf of Dr. Oparin and then says Dr. Oparin would prefer to have me present him with my questions in writing, to which he will then respond by writing me in America. Well, of course, that would be fine, I reply, but it would be helpful if he could answer a few of my questions informally while I'm here. After all, I have come a great distance . . .

Dr. Oparin mulls it over. There is a brief exchange between the two men. Oparin gives a shrug and then nods. "Dr. Oparin will give to some of your questions an oral answer," the translator says. "Do you have some copy of them?"

As it happens I do have them in writing, but only one copy. I hand it to him; he studies it closely and then, after a prolonged and somewhat awkward silence, delivers what I take to be a translation of all my questions to Oparin, who in turn thinks a bit and then rumbles forward, pursued by the halting English phrases of his young assistant.

"Number 1. At the beginning of the century the origin of life theory was in great crisis . . ."

With alarm I realize he is giving his answer without first reading the question aloud from my paper. (For a wild moment I am reminded of Steve Allen's old Answer Man routine—"The answer is, 'Dr. Livingston, I presume.' What is the question? The question is, 'What is your full name, Doctor?' ")

"Only the idea that life is the result of the evolution of carbonaceous substances gives a way to avoid this crisis," the young man goes on. "The developments leading to the origins of life can be divided into two parts, or steps. The chemical step and the biological step. To provide the concepts of the chemical step, the more important investigators are astronomers, physicists, chemists, and so on. The concept of the biological step is based on investigations in the fields of chemistry, biochemistry, and paleontology."

Would the translator repeat the question before proceeding with the answer? The young man looks at me, then to Oparin, who shrugs and nods.

"Question 5." (Oparin is evidently choosing only the questions that suit him.) " 'Will it ever be possible to pinpoint in time the origin of life?'

"Dr. Oparin says the investigations of paleontologists tell us that the process of the origin of life has a duration of more than three billion years. Many paleontologists say the origins of life took place more than three and a half or four billion years ago. Dr. Oparin considers that this point of view must be proved more carefully. He thinks that life originated only three billion years ago, or perhaps slightly earlier."

There is a pause. We all look at each other somewhat tentatively.

"Question 6. 'Why does the fossil record tell us so little about this question?'

"Dr. Oparin says that more than two and a half billion years ago we had a great number of fossils that have a biological origin. The fossils that have an age of more than three billion years may be of biological origin, or maybe they are organisms which are predecessors of biological organisms.

"Question 8 . . ."

(Question 7 was: "Is it accurate to say the chemical emergence of life took longer than the subsequent development of organisms?" Evidently he feels he has already answered it.)

"Question 8," he goes on, " 'If we can say that the main difference between living and nonliving is that the former has a sense of 'purposiveness,' how and when can chemical components be said to have gained purposiveness?' "

The old man is interested by this one. He speaks at great length before the translator resumes. Then he watches me closely. "Dr. Oparin says that chemical evolution leads to more and more complex molecules. Then the final step gives purposiveness to the very complex molecules of protein and nucleic acids. Then these complex molecules form a self-constructive organism that can be considered as living. This purposiveness concept is a product of the old mechanistic conceptions that see a parallel between organisms and machines. When machines are constructed, the parts that are made with a definite purpose are connected, and so the machine is created. But the purposiveness of the construction of these parts is based on the previous plan, and so it has a beginning in the creating action of the designer. We can't suggest something like this when we speak of the origin of life." Oparin resumes, going on at some length, ignoring the translator, and now he adds gestures. The translator gathers himself.

"*What is this purposiveness?* It is the connection between the form of the part and the function of the part! For the eye in itself, the function of seeing has no sense. It has sense only when the whole organism is taken into consideration. So the function of the catalysts for the protein molecule or the function of the genetic code for the nucleic acid have no sense when considered apart from the organism in which they occur.

"Purposiveness appears only when the whole phase-separated system appears. In the scientific literature, there has taken place a discussion of which had the earlier origins—the protein molecule or the nucleic acid molecule? Something like, What had the more early origin, the egg or the hen? Nor can molecules which possess nucleic acid possess the functions that they will have *after* the organism appears. The purposiveness can appear only as a *result* of the natural selection of the multimolecular phase-separated systems that can undergo natural selection connected with the environment. So Dr. Oparin takes a position opposite to the concept of reductionism. The parts do not define the properties of the whole, but the whole in its development defines the properties of its parts."

Oparin looks sharply at me as though to see if I have understood. I understand only that there is no satisfactory answer to the question I have asked, and so I nod.

"Question 9. 'Explain why life could arise only in the hostile environment of early life, and not on the present-day earth.'

"Dr. Oparin says that oxygen is not necessary for life to begin. We know that the primitive bacteria do not require oxygen. Oxygen can damage them. The data of paleontologists show that organisms requiring oxygen appeared only after organisms that did not. Why couldn't life arise today? The answer is a little bit paradoxical. Life cannot arise now because it has already appeared." Oparin smiles broadly. "If primitive organic matter appears it will be eaten by living organisms.

"Question 10. 'What are the most important of the chemical experiments that have taken place since Stanley Miller's original experiment?'

"Dr. Oparin says there are a great number of experiments taking place in Russia and the United States, experiments of Ponnamperuma and others." He does not elaborate. Oparin whispers to the deputy director, who rises and leaves the room.

"Question 11. 'Will these experiments ever be able to re-create a credible sequence for the origin of life?' "

The translator waits while Oparin thinks about this. "It is a great interval between these experiments and the most primitive living organisms. Dr. Oparin is sure the results of these experiments will be the creation of primitive living cells in the test tube. But there is still a great difference between a living model and living organisms created as a result of evolution."

Question 12, having to do with whether this knowledge, once gained, will change our conception of ourselves, is also ignored.

"Question 13. 'Doesn't reductionism rob us of our spiritual possibilities?'

"Dr. Oparin says the most detailed investigations of the processes in modern living organisms cannot lead to deeper understanding, to a whole understanding of life. We cannot understand the sense of purposiveness based only on experiments of modern biological processes. Reductionism is valuable for biological studies, for medicine, and for the practical study of life. But when the reductionist deals with the problem of the *origin* of life, he can do nothing."

I ask the translator if I may ask a direct question of Dr. Oparin. The request is relayed; there is a measuring look and a quick nod.

"Would you comment on this quote from Salvador Luria of Cornell? 'In order for life to evolve, a form had to come into being that could direct its own replication.' "

Oparin resumes, followed, after a brief interval, by his translator. "The origin of organisms having a modern replication system based on the genetic code is the result of the primary evolution of the more primitive organisms, and this evolution can have a sense of Darwinian evolution in its, uh, first sense—the elimination of organisms that are less fit for survival. Now in our laboratory we get models of these primary phase-separated systems, the so-called coacervate drops. They have an ability to interact with the environment, taking matter and energy from it. So they can grow and divide into daughter drops, and give a set of new drops . . ."

Does this mean Oparin believes he has developed a self-replicating system? But that is all I am going to get. The translator indicates Oparin has said what he will say. He will write me at greater length in New York. Now, as though on cue, the deputy director enters with a tray bearing a bottle of brandy, cookies, and candy. Oparin toasts the success of my trip, and he and the deputy director throw down their brandy at a gulp, without blinking. It is the last I will see or hear of him, except for a greeting card at Christmas, a year later.

3

THE FIRST DAY

Trout Lake, Washington

In the darkness of Sunday morning, the only sound that of the stream somewhere out to his left, Vihtelic sifted his priorities. At daylight the traffic would start and he needed some way to let the drivers know he was down here. He had to find some way to free his foot, and he was sure he would. But what could he tell them at Drake Willock? What could he do about what he'd done to their car? The answer was, nothing. He would lose his job. How could he expect them to understand? That was something to worry about and it had worried him more than anything else through the long night, as it worried him now, but there was nothing he could *do* about it. He had to start thinking about what he could do something about. His foot hurt now, even when he didn't move. That was trouble he had to do something about as quickly as possible.

Things he had learned in the army about disaster situations began to come back to him. The first rule was to remain calm. Even before checking a patient's medical condition—even if he was spurting blood —you had to remember first of all to stay calm. The next thing after that was to get organized. He remembered the five *P*'s—Prior Planning Prevents Piss-poor Performance. Maybe somebody was looking for him right now. He yelled, "Where are you?" but he could hardly hear his own voice above the steady rumble of the unseen water. Whether there was a God or not was anybody's guess, but he didn't worry the question now—he prayed. No Hail Marys or special promises for special favors. Just a straightforward direct request for help. He prayed out loud: "Please God help me get out of this car so I can live."

Thump-*thump,* thump-*thump.* A car was going over the wooden bridge. Vihtelic screamed and twisted himself toward the window at his left. The sound of the stream was too loud; how could anyone hear him down here? He waited through the darkness, and when first light came he took his time mapping in his mind what now he could begin to see. The stream was about ten feet wide, about fourteen feet from the near bank to the wagon, rushing down the center of the ravine which itself was very narrow, slender, an envelope in a mountain forest, about thirty-five feet across at the top—really just a vertical slit in the woods. He couldn't see where he'd come down, but he could look across the stream and up the far bank to where the trees stopped and the road must be, a sort of thinned-out edge running across the thicker foliage to the left and then blocked by the wagon's window frame. Very green, very lush, very very lush. Tall pines; some of them had fallen and washed down the side of the ravine. Where he was was like a small riverbed at the bottom of a green slit in the mountain. Lush above, rocky down here. Wilderness. He wondered how long since anyone had been down here. There was no rubbish—no beer cans or Styrofoam cups, nothing. Years, probably.

What he saw about him now formed the special conditions for the plans he would have to make. He wished he could see more of the road. What he had to work with were the pieces of the car he could tear loose and those of his things he could reach. Thump-*thump,* thump-*thump.* Another car! Vihtelic yelled again and turned toward the window until the pain from his foot checked his movement. The car moved on. He waited, and after some time he heard another one passing over the bridge. This time he made himself lie still and watch for its movements. For a few seconds he could see the top of it passing along the edge in the foliage above, and for just a second, no more, he could see the whole frame. Any driver glancing down then would be able to see him, that was for sure. So it was just a question of waiting it out. But while he was waiting he'd better get organized.

Organizing your situation properly was the first step toward forming an escape plan. He remembered training films he had seen about prisoners during the Korean War. Those who survived were those who thought about escaping and worked at it, whether they got out or not. The Turks and the Australians. Those who did nothing died. The training film called it "give-up-itis." Vihtelic decided he would build his plan toward the objective of helping people find out where he was—not that anyone had missed him yet—but *anybody,* which today meant most likely a passing motorist. The first part of the plan would be to find a way to help people see he was down here. The second part of his plan would be to work at getting his foot loose.

Throughout the early morning, the pain had steadily worsened. Most of his effort in thinking straight now came from trying to get his mind to focus on anything other than the pain in his foot. Because of the constant pressure from his toes pulling against the tendons, there was no way he could position his body within the wreckage to relieve it. Earlier, around midnight, he had worked his way through a series of contortions that served him now as routine. Buckling up beneath him, a fold in the roof of the car was pivot point. He could move the trunk of his body only to the right of it, twisting his hips counterclockwise. Bending his head and pushing up against the seat, he could rise halfway back up on his knees. He could cross his arms over his chest and rest his head on his hands, as in the fetal position. But no single position promised him relief from his pain. Having to move, he moved with the dread that every movement aggravated it. But there was nothing he could *do* about it except get out.

He began an inventory of his resources, all of which were within the battered car that imprisoned him or were scattered on the ground about it. He found his tennis racket in the back and next he found a tire iron which had been thrown forward near the seat. It was L-shaped with a lug socket on the short end and a beveled edge on the other. The air mattress and the other sleeping bag were some distance from the wagon. The cooler was closer but smashed in and without any value that he could see. Where was the coffee pot? He could use that when it came to water, which he would soon enough have to think about, but it was missing. There was a roll of masking tape in the glove compartment. He found the metal hot-dog sticks, some more clothes, the camera. His suitcase was in the rear of the wagon. But he couldn't find the paring knife or his toiletry kit. And, of course, he had the wagon, or those parts of it he could reach, as a resource still to be explored.

Vihtelic pushed up against the seat and worked the wedge of the lug wrench behind him into the first opening between the tree root and the dashboard, pushing down and then up against the weight of the car while pulling on his foot. The pain in response to the effort was blinding. It rose up through his leg across his groin into his penis and up through his chest: a flash shifting to a steadily penetrating, deepening pain. There was no movement of the wagon in either direction. He rested and wedged again, working on despite the pain until he was exhausted. Another car rumbled over the bridge. He banged on the wagon frame with the lug wrench, screaming at the top of his lungs. The car drove on. Vihtelic was scared to death. He wasn't going to bullshit himself about that. He cried for a few minutes but then he made himself stop. He wasn't going to give in to that.

The Archeozoic

At some time within the first billion years of the earth's history, a membrane wall formed about the two large molecules of nucleic acid and protein, shutting in from the outside their facility for self-replication. Once enclosed, the several parts became an identifiable entity, a whole: a cell, as in the literal sense of the word—a "little room." The cell was the first clearly defined, structurally organized manifestation of life. Microfossil evidence of this entity, dividing in cinematic progression from one part into two, has been found in the Onerwacht rocks of South Africa, 3.2 billion years old. Within the early primordial seas, other less well-defined organisms had formed as well, and all these primitive things, including the first cells, existed by feeding on organic molecules, dinner and diners alike the products of lightning in the primordial soup.

At some unknown time within the first billion years, the growing population of organisms exceeded its food supply, endangering the future existence of the cell. Survival called for yet another innovation. As ingeniously as the first cell had evolved its own replication, a kind of cell now developed a fueling system based on solar energy. By using the sun's light to act upon chemicals within its own walls, splitting the structure of water to form oxygen, this cell began to produce its own food—*within itself.* From the cell which had fed outside itself had now come self-feeders. (From surviving outside feeders eventually would evolve all animals. From self-feeders would evolve all plants.) In effect the self-feeding cell had internalized the phenomenon of lightning in the primordial soup.

This event changed more than the internal structure of some cells. As such cells multiplied, their new fuel system—called photosynthesis—produced oxygen in such abundance that it eventually rose into the atmosphere and accumulated about the earth. An old enemy became a new ally. Whereas life could not have begun in the presence of oxygen, its presence in the atmosphere became an essential condition for the development of life yet to come. From the chemical conversion of solar energy through the cellular process of photosynthesis, green plants of the future would produce the oxygen animals would need to live, and animals would produce the carbon dioxide plants would need to live. Binding the process into an irreversible need, oxygen formed about the earth a shield against the effects of ultraviolet radiation which otherwise would have prevented colonization of the land.

For animals oxygen would provide a powerful source of energy; the larger and more complex the organism, the more of it would be needed. For example, to serve their complex purpose the cells of the human brain draw ten times more oxygen than other body cells. Thinking about this—the very act of thinking about it, as René Dubos has observed—is the long-range but nonetheless direct consequence of the development of oxygen power by the cell through photosynthesis.

After the passage of two and a half billion years of the earth's history there came a new development of almost equal magnitude. Inside its wall the cell formed yet another wall—an inner membrane encasing the molecules of nucleic acid and protein. Within the little room there had formed an even smaller room, a nucleus further insulating the cell's vital parts from the outside world. This, in its new, nucleated form, was the first modern cell. The processes of sexual reproduction, in which genetic information is exchanged and recombined, came with the nucleated cell, and so did all the possibilities of diversity and change. Descendants of the first nucleated cells now cover the earth: they come in various sizes and impressive quantities, and they perform remarkable feats. A bacterial cell (nonnucleated) may measure one-tenth of a micron, which is 10^{-4} centimeters. An average human cell is ten microns across. A single human nerve cell, while quite small in diameter may measure several feet in length. There are a hundred billion (10^{11}) cells in the human brain, consisting of 10^{15} large molecules. (As an indication of the magnitude of this number, 10^{15} minutes represents the amount of time that has passed since the forming of the galaxy which produced the sun and the earth.) At first, nucleated cells collected into colonies in which all cells functioned similarly. In time, certain groups of cells began to develop specialized functions (such as light perception) leading eventually to the evolution of complex organ systems.

The early emerging, fully formed nucleated cell was a blueprint for the future of life, and the order proceeding from it is consistent with the plan it established. The distinction between the cell without a nucleus and the cell with a nucleus is the most important division in the description of life on earth, more important than the later division between plants and animals which, in ways beyond sharing nucleated cells, are actually more alike than unalike.

Over the first nine-tenths of the world's history these several developments occurred on the microscopic level within the seas. However distantly obscured their workings, it is now clear the first nucleated cell arising a billion and a half years ago and flourishing through the beneficence of photosynthesis marked the true beginning of the richness of earth life, the basic organizing unit from which all else would follow: weeds and trees, whales and men.

College Park, Maryland

We could have lunch when I came down, Cyril Ponnamperuma had said over the phone, and perhaps I might want to stay on to see his laboratories: one for biochemistry, and still others for geochemistry, organic chemistry, and planetary atmospheres, including "clean rooms" for working with moon rocks, meteorites, and ancient rocks of the earth—all part of the Laboratory of Chemical Evolution, which he directs, at the University of Maryland in College Park, the last stop on Amtrak out of New York before reaching Washington. He would be pleased to take up where Oparin left off.

Beyond the elevator opening to the third floor of the chemistry building—the university has the largest chemistry department in the country—the way to Cyril Ponnamperuma's office is marked by Faustian icons: photographs of the latest in primordial soup machines; of the Eastern Seaboard of the United States, from the vantage of a satellite; and of the planet earth, from the vantage of the moon. A hand-lettered sign reads: *The Water You Drink Can Cause Cancer.* On a bookcase inside his office there is a can of Campbell's soup with the word "primordial" pasted over "tomato," and beside it, on a mounted stand, is a fragment of the Murchison meteorite which fell to earth in Australia on September 28, 1969, bearing the first tangible evidence, in the form of amino acids, of the presence of chemical evolution off the earth, somewhere else in the universe. A framed drawing on the wall shows Ponnamperuma in his laboratory extracting from a test tube a smiling, voluptuous nude while being observed from behind by a portrait painting of A. I. Oparin, smiling too. The setting for it all—motel modern with orange acrylic rug—is aggressively cozy and informal.

We will have crab cakes at the university restaurant across the campus; it is walking distance from his office, just beyond a practice football field across the way. As we head in that direction, Ponnamperuma turns immediately to the aging Russian who had left me hanging. Coacervate drops, I say, were his last words.

"Yes, well, Oparin's idea was simply that the first organic compounds were formed from the energy available. But Oparin in his laboratory never once worked with organic molecules. He left it to Miller and Urey to do that experiment. For the past thirty years, Oparin has dealt with the problem of *organization,* how things come together—with coacervate drops." Coacervate, from the Latin word *coacervare,* means, I had learned, to assemble, to cluster. "You see," Ponnamperuma continues, "we would never have had life unless the nucleic acids and proteins interacted. But how could they if they stayed miles apart? Even if the primordial soup is full of organic compounds there's still a big jump between the organics, the molecules, and a cell. So Oparin is after a way of bringing these prebiological components together. Of all the ideas put forward, so far his are the most logical, the most scientific."

Thus Oparin's reference to daughter droplets, Ponnamperuma says, has to do with the rendering of precisely this problem—how to show the convening of organic compounds in an assemblage that would permit life to form. The movement together would come through electrical attraction of the several parts, and in the process the water separating them would be "excluded"—a difficult feat, as one researcher has observed, rather like a swimmer trying to stay dry in the water. "Anything that promotes the removal of water can enhance coacervation," Ponnamperuma said. "If particles are electrically charged, opposite charges will attract each other and the particles can collect, ultimately to form drops, water being excluded in the process." Charged molecules were present in the primordial sea. The adsorption of water on the surface of a coacervate could suggest the beginning of a cell membrane and thus the beginning of individuality.

Since primitive organisms no longer exist, Oparin's experiments try to replicate the earliest sequence as an analogue, which goes roughly as follows: droplets floated about in the primitive sea, some of them with special chemistries. Over time—long reaches of time—some droplets developed the ability to take molecules from their surroundings and alter them to produce substances that ensured their survival, not only as parent droplets, but also as daughter droplets into which they eventually divided. (Soap and oil globules behave in this fashion.) This may have been within the moments when life began, but if so, how exactly did the droplets manage to reproduce themselves? And how did genes form to become the basis of heredity, and proteins that would allow the droplets to grow?

"We are studying relationships," Ponnamperuma says. "You see, the code is a universal one, whether you are looking at the smallest microbe or at man. There is a particular relationship between protein and amino acid in all living things. Why do we have that? Because there must have been a relationship between these two *before* life. And we are trying to discover it. Once we know *that,* we can reconstruct it, we can go back from what we know of life today. Then we are not far from making a replicating system in the laboratory."

"How far?"

The hostess has led us to a table in a corner. Ponnamperuma motions me to a seat across from him. He smiles. "That depends on how bright my graduate students are. Five, ten years. Maybe even closer. We have the idea, and we're not completely blind. There are various ways of going at it. The problem is complex." He speaks as though half to himself, urging himself on. "We are going from scratch—we want to get every step." He shrugs, smiles again, and turns his attention to the menu.

Ponnamperuma comes from Sri Lanka. He is the youngest of three brothers, all chemists. At the University of Madras he studied under Jesuits the religion of the Hindus, Buddhists, and Christians, and then went on to the University of London to study chemistry, which he thought offered a better opportunity to make a living. There he attended a lecture on the origins of life by J. D. Bernal, the physicist who had independently arrived at the same conclusions as Oparin—though five years later and in another language. As Bernal discussed the prevailing opinions scientists held then, in the late 1940s, Ponnamperuma realized that his two quite disparate interests were intersecting. The connection beguiled him then and does still, to the extent that he is willing to go to the mat with Creationists, fundamentalist advocates of the Bible's version of genesis, on their own terms. To his lectures, where he is sometimes questioned by them, he brings his evidence in the form of microfossils he has found in ancient rocks.

"They say to me, your logic is very good but how do you reconcile your version with the story of creation? And my immediate answer to them is, *which* story of creation? I come from a society where six hundred million people think of matter as being eternal. In the Hindu and Buddhist philosophy, God is everything. We are a part of God. In the Western concept, God is a being who *did* something. The idea of God depends a lot on your background. The burden is on the philosopher, not the scientist. We are not trying to disprove anything. We're trying to inquire, to understand, and this does not take away from the spiritual nature of man. On the contrary it gives us a sense of our place in the universe. We become humble on the one hand and on the other we are exhilarated." The waitress is hanging back, waiting for him to finish. "We are brothers and sisters of the stars!" Ponnamperuma returns his attention to the menu and orders crab cakes for both of us. He has raised a possibility I am as sensitive to as anyone else, and for a moment I reflect on some of the things I've read before coming to see him.

If it *is* true that the metals found in trace amounts within the human anatomy may be tracked backward to the explosions of stars existing before the sun appeared, and before that to the primordial fireball of twenty million years ago, its cool echo still speckling the television screen in the playroom, no theology seems quite adequate to encompass such dimensions of time, space, and chance. And without some minimally satisfying account of man's appearance on earth, religions must falter, faith give way to reason, mere ethics take the place of holy law. And yet (I read), because man at his best is purposeful and because there is an orderliness to the systems of the universe, the belief that a purposeful plan lies beyond man's present grasp—beyond even his ability to imagine it—persists in the thinking of some scientists.

One astronomer seeks to relate the Big Bang and its subsequent sequences to the Bible's seven days of creation. Even beyond the metaphor, some biblical parallels are striking in their anticipation of later scientific discovery. "I also am formed out of the clay," Job declares, and Salvador Luria quotes him in remarking on how clay, present on the primitive earth, assisted in the early linkage of amino acids in precellular structures. (Indeed, clays in which nickel is present adsorb among many possible amino acids only the twenty used for life's proteins.) Yet such versions of genesis are necessarily anthropocentric—man-centered—and too restrictive even as metaphors for most scientists. "I don't believe in the God of man," Einstein once answered a reporter, "I believe in the God of order." In 1911, George Bernard Shaw, seeking to keep abreast of science's discoveries, posited an evolutionary God, learning by trial and error as we—and He—go along. The tragic failure of man to realize his own potential is man's problem, not God's. If man fails finally in God's terms, he will simply cease to exist, as have other unsuccessful species throughout known history.

In any case, the biochemist Ponnamperuma seems to have no trouble with the spiritual implications of his own quest; for some time now he has pursued it with exuberance. He decides to take a moment and tell me about it.

After university, his first job was with Melvin Calvin, the biochemist who used carbon dioxide rather than methane and ammonia in his recipe for primordial soup. When the National Aeronautics and Space Administration asked Calvin to suggest a scientist to work on extraterrestrial life, Calvin recommended Ponnamperuma. On the second day in his new job Ponnamperuma was asked to set up a laboratory for the study of the origin of life, and he has been chasing after it ever since. Why was NASA interested? For several reasons, Ponnamperuma says. Understanding how life came about might help to explain how it could arise elsewhere. Moreover, any extended space voyage (as with Gerard O'Neill's space colonies intended to orbit an eroded, abandoned earth) must be self-sustaining: once earth is left behind, food, air, and water would have to be produced through regenerative processes. Life processes, as derived from the physical and chemical, would have to be made to work—and keep on working—within the confines of a space ship.

At NASA, where astronomy and physics formed the estate of the privileged, Ponnamperuma and his associates were not taken very seriously until 1969, when hydrogen cyanide and ammonia were discovered to be present in the interstellar medium, the significance of which was to suggest the strong possibility of life occurring elsewhere as it has here. Just after that event, Ponnamperuma recalls, "during the first day of a conference I was holding, the astronomers sat on one side of the table, the chemists on the other. By the middle of the second day they had begun to intermingle and connect." At NASA, astronomy, physics, chemistry, and biology began to converge.

Such connections did not lead to unanimity among all scientists, however, and I mentioned to Ponnamperuma Martin Rees's reservations about the probability of life existing elsewhere, and as well the resistance of the British philosopher Sir Karl Popper to the suggestion there has been an acceptable explanation even for the emergence of life on earth (". . . a riddle," Popper has written, "which turns back on itself").

Ponnamperuma says crisply, his lunch growing cold, "If the laws of chemistry and physics are universal in character, the same sequence of events should occur elsewhere in the universe. About 94 percent of living matter is made up of hydrogen, carbon, nitrogen, and oxygen. These are the very elements which are most abundant in the cosmos. The universe is a very scientifically organized entity—there's very little *chance* there.

"On earth all life has amino acids, proteins, and nucleic acids. Twenty common amino acids, nucleic acids with four bases, two sugars. The alphabet of life is twenty-eight letters. When you go back and look at the interstellar molecules you see very clearly why this is so. There is an inherent propensity to use these molecules in the universe.

"Hydrogen, carbon, and nitrogen give you hydrogen cyanide. Hydrogen, carbon, and oxygen give you formaldehyde. If you have those you get amino acids. If you have hydrogen cyanide you get the bases. If you have formaldehyde you get the sugars. So there is a pattern—the distribution of electrons around individual atoms."

One senses that this would still not be so conclusive an argument for Rees. "Are there as likely to be planets with atmospheres similar to ours?"

"Other planets were formed in the same way. The processes of atmospheric evolution are the same."

"Different constituents wouldn't form different systems?"

"No. The primordial nebula of any solar system would have to be the same. If somebody teaches the periodic table on a planet around another star, he would be teaching the same thing I teach my students."

Ponnamperuma prepares to indulge his sweet tooth with a napoleon. I sorely miss the presence of Rees. "But if environment conditions the shapes of life, can you so quickly assume there will be humanlike creatures in other places to communicate with?"

"The answer to that would be that the chemistry of life is the same—life is an extension of biochemistry. I think you'll have life only in environments similar to earth, of course. But the processes are the same, the differences within them would be minimal. I wouldn't be surprised if you landed on some planet like earth and somebody about five feet two, with two eyes, came up to you and said hello."

Ponnamperuma finishes his second cup of coffee and inquires as to whether there is anything else, at this moment, I should care to ask. Yes, I say, and I bring forth a quote I have in my notebook from a prominent geneticist on the consequences of invading the life processes at their most private recess: "My generation," I read to Ponnamperuma, "has been the first to engage, under the leadership of the exact sciences, in a destructive colonial warfare against nature. The future will curse us for it."

Ponnamperuma is untroubled. "Evolution is under our control now. We can direct it. Take petrochemicals—oil. In our work we have come up with the idea that maybe the primitive oceans were *all* covered with oil. Maybe this primordial oil slick was so vast it would make Santa Barbara fade to insignificance. When you think of the amount of hydrocarbons that could be generated! So there is *that* side of it. We might have come to the point where, because of the oxidizing atmosphere now, this process has been retarded. But if you can control such things—if you have such processes in the laboratory, if what I'm doing in my little flask I can do on a giant scale, if I could come up with a way to *regenerate oil!* . . .

"Or take food. We get our amino acids from vegetables and animals. Now we can make amino acids from primitive atmospheres. We might come to a time when we can make them directly from hydrogen, carbon, and nitrogen. We might be able to synthesize *food!* NASA is interested in supplying food in the long term, perhaps over several years, to astronauts. You can carry your protein with you in tablets, but to carry carbohydrates is too large a problem—too much bulk. But, you know, each astronaut gives out a pound of carbon dioxide a day. Maybe you can work that carbon dioxide into formaldehyde. You see, there is hydrogen given out from all the fuel cells, and it is wasted. If you combine these two, from formaldehyde, you can make sugars. You may have to feed them on treacle but you can provide carbohydrate. Some of the NASA people have come up with a very worthwhile project to solve the starvation problem on earth. Everything you burn gives off carbon dioxide, and instead of that carbon dioxide going to the atmosphere, let's convert it to food. Into bread! We are not far away—we know the principle of the reaction. I expose formaldehyde in dilute solution to ultraviolet light or to an energy source and I can get the sugar. I must come up with methods of separation and so on . . ." He looks at me to see if I perceive what he is saying. "But those," he says, almost scornfully, "are only *technical* things."

4

WATER

Trout Lake, Washington

Vihtelic wasn't hungry. Except for the apple he'd eaten the first night and a package of cheese crackers he'd found and tossed aside because they were wet as the rolls, the thought of food had scarcely entered his mind. He was mildly surprised to discover he really didn't care that the rest of his food supplies—eleven apples, the wheat bread—were missing. He assumed they must be close by. Last night field mice had got into the wagon and run up and down his legs and his chest so long as he didn't move, which was no more often than he had to and never because of the mice or the ants crawling over him. Moving brought pain, but he couldn't sleep more than a few minutes at a time because his body would cramp. So he had to move. The mice and the ants were a very small problem. They were after the food, which held little interest for Vihtelic.

Allowing for the proper temperament, a man in good physical condition can go a long time without food. Except for his foot, which now he feared he might lose unless he got out soon, Vihtelic was in excellent shape. He was six feet tall, weighed 190, and he'd worked at keeping himself in condition, jogging regularly and cycling up to two hundred miles on weekends. These were plusses. Another plus was that Vihtelic knew more about the needs of his body than would most people unfortunate enough to find themselves in such circumstances. He'd had a year as a medic in the Green Berets and two semesters in biochemistry at Hope College after that. Deprived of food the body burns protein reserves drawn off its own fat. He knew this much. After the fat goes it draws from muscle. As much muscle as you have on you, that's how much food you have inside you to stay alive. Amino acids and protein—that's what muscle is made of.

But he knew, too, that he couldn't last very long without water—three days maybe, at most, before the consequences of dehydration set in. The first warning signs were hallucinations—half dreams, as in the last moments of consciousness before sleep begins. Toward the end breathing becomes rapid—the body seeking to compensate for fluid loss, to hold the acidity level in the blood by blowing off carbon dioxide. Some of this Vihtelic knew, but very little of it—beyond the pressing fact that he needed to get water right away—did he want to think about. He would be out of here before he was in real trouble. That was for sure.

Today was when they'd come. He would be missed at the office and everybody there knew he had gone hiking at Rainier. They would call park headquarters and very soon people would be looking for him. But this didn't mean he should just lie here and wait. He had to get water, and he had to find some way to signal to his rescuers that he was down here. One way or another he was going to get out. He'd read that book about the rugby players stranded in the Andes after an air crash, and the one he had identified with was Canessa, the medical student who had kept the others going. Canessa had his shit together. Reading about him, Vihtelic had thought at the time that was the way he would have been. He would have got out, no matter what.

Now such a thing had happened to him, and he was confident he could meet the challenge. He had to establish a routine, and the first thing he had to do today was to make himself as neat as possible. He placed himself to urinate between the frame of the wagon and the door next to him. He had his comb in his pocket, so he combed his hair and tucked in his shirttail, the most he could accomplish without water or his toiletry kit. He wound his watch. Then he turned his attention to the parts of the wagon he could reach. Now he would inventory his resources.

There was wire under the dashboard. He pulled out a good bit of it and stockpiled it beside him. Scanning the roof beneath him he noticed that the cloth fabric was held in place by rods, metal probably.

There were six of them. Vihtelic twisted them loose. Among the debris in the back was a Coke can. With the wire and the rods, and with the Coke can as receptacle, the thought occurred to him that he might have a way to get water. He could fish for it. He hooked the ends of the metal rods and fixed them together with the masking tape. He knocked holes in the top of the Coke can with his tire iron and raveled the wires into a single, long strand. On one end he tied the can and on the other the rod pole.

A large, half-sunken log lay between the stream and the wagon. He would have to lob the can some fifteen or sixteen feet, out and beyond the log, far enough into the stream for the can to sink. Bracing himself against the window frame, Vihtelic began to throw. At first he had little control over the direction of the can, but he was patient and his aim began to improve. There was a definite rhythm to the sequence, and he worked out a chant to keep himself at it. Out loud, he urged himself on:

Seven wraps of the string,
Throw as hard as you can,
Lots of loft and
Over the log!

Throwing it out over and over, he worked at developing his skill. He was fishing and the water was his trophy. Eventually the can hit its mark, sinking slowly into the clear water. Vihtelic eased it in, working it carefully over the log. He drank the water and threw it back. He drank a second canful and threw it back again. The wire snagged on the log. When he tugged at it, the wire broke. He found a beer bottle in the back of the wagon, tied it to the pole and started again: "*Seven wraps of the string . . .* " He threw it out too hard and the bottle broke. He found a Coke bottle, but that broke, too. Vihtelic cried.

Not for long. He made himself cut it out, and he turned back to his foot, defying the pain and wedging the lug wrench against the dashboard, up and down. To offset the pain he would think about other things. He shouldn't expect a lot of traffic today, Monday. The tourists had gone home. Only his rescuers could be counted on, the search party he was sure by now had formed and was tracking his last known movements. How would they know where to look? Vihtelic thought back to Saturday. Just before starting up Rainier he had called Mary from a pay phone, but he hadn't told her he was going on to Hood. How could he? He hadn't known then that was what he would do. He stopped wedging the wrench; it hurt too much. He'd met up with a sailor near the end of the hike, a guy on weekend liberty, and they had come down the mountain together. A guy named Gary, or was it Brian? He couldn't remember now, nor could he remember telling him his late-formed plan. A park official had logged in the time when he had returned his pack and hiking boots. It was in the park's records that he had left Rainier on Saturday afternoon, but there was no record of his next destination. The old-timer at the tavern—*he* would know. If the rescue party asked around Randle, somebody would remember his coming through asking directions for a shortcut.

Of course it would take some time for them to narrow their search, Vihtelic realized. He would have to understand this. His brothers would get things going—they were probably already out here. He just hoped his mother and Mary could understand, too, and not worry too much. Mary would be all right, he was sure, but his mother was overweight and nervous and he feared for the anxiety he might be causing her. It would be the same if any one of them was lost but Vihtelic didn't like being the one to cause her concern. He didn't worry so much about his job anymore. That was done, and there was nothing now he could do about it. He did have to worry about saving his foot, however, because there was no way he could put the urgency of it aside. Vihtelic didn't know how long before the loss of circulation would bring gangrene. Not knowing, he could accomplish very little by thinking about that, but he fashioned a probe from a long sliver of broken glass beneath him and poked at it anyway, hoping for some local sensation. It hurt all over as it had since the pain began but there was no feeling from the glass. He pushed the glass shard into the flesh. Nothing.

He had better turn back to finding another way to get water . . .

The Proterozoic

Moving together, multiplying, dividing, and moving together again, nucleated cells began to evolve into multicellular creatures of all sorts, of different sizes and odd shapes. The consequences of their gatherings may be seen today in the evidence at hand. There are three million many-celled species alive now, maybe more, survivors in a harsh world which over the past billion years has seen many others come and go. All their ancestries trace to the nucleated cell, the first progenitor, founder of the kingdom, as it has since come to be called, of Protista. From Protista three more discrete kingdoms have emerged which include all forms of life thereafter: the kingdom of plants, the kingdom of fungi, and the kingdom of animals.

The evidence for the rise of each of these lesser kingdoms is piecemeal and circumstantial, parts of it coming from the microfossils of ancient rocks, and from comparative analysis of living things; but most of it lost to the seas within which they formed. Although the mushroom and its kin are fungi, which might be (and have been) mistaken for plants, the division between plants and animals would seem clear enough. In marking the division, however, outward appearances are deceptive.

While some differences are obvious—plants have hard cell walls, animals have a thin cell membrane—plants and animals share a common heritage of biochemistry, genetics, metabolism, and intracellular organization. Plants are the chief food source for both animals and plants. A way of distinguishing among the three kingdoms has to do with the ways in which each derives its energy. Plants draw on the sun to create living matter through photosynthesis; fungi ingest organic substances; animals ingest plants and/or other animals and sometimes fungi.

The exact time of origin of each of the kingdoms remains unclear. An early representative of the plant kingdom was green algae; of the fungi kingdom, slime mold; and of the animal kingdom, the sponge. The sponge is notable beyond its category. Until a hundred years ago there was uncertainty as to whether it was an animal or a plant. It is hard to get rid of a sponge. Crushing it will not destroy it. Sift its cells through silk mesh under water, and they reconvene into a mass from which new sponges will emerge. Although it is considered the simplest form of all many-celled animals, it poses a perplexing question: what exactly constitutes a living individual?

Less is known of the origins of the kingdom of plants than of the kingdom of animals. The earliest of plant ancestors evolved from green algae, although how and when are questions still to be answered. Some seaweeds are algae. Sexual reproduction began first in plants. At the edges of new land forming out of old seas, plants began to spread and grow, probably about 450 million years ago. To solve the problems posed by gravity, they developed stems, and later wood, for support (as animals were to develop skeletons); to solve the problem of obtaining water they sent roots into the earth; and to discourage its evaporation, they developed hard surfaces (as did animals). After plants came ashore animals followed, obtaining protection from the killing force of the sun from the ozone which began as oxygen produced by plants, breathing the oxygen they provided, and eating them to obtain energy.

Fossil fuels are so called because they come from early living things long gone: oil from the compressed remains of plants and animals within the early seas; coal from the remains of plants. Three hundred million years ago, the great carboniferous forests spread across the land in all directions. Horsetails the height of trees, giant ferns, and ground pines formed these forests. Through the slow, steamy-wet period of their decay, lasting over 125 million years, the world above water was a drab place, of greens and browns, and it would remain so until the angiosperms, or flowering plants, arrived—an event thus far unexplained, and of immeasurable consequence.

Beech, birch, maple, walnut, fig, oak, magnolia, poplar, willow trees, vines, holly, ivy, giant redwoods climbing 350 feet into the air, fruit trees, vegetables, wheat—all come from the angiosperm, the most successful of all plants, appearing in abundance by 100 million years ago. Like the nucleated cell, the angiosperm managed to encapsulate its essence within a protective covering, its seeds within an ovary. This is what made it special. With the arrival of angiosperms, blossoms appeared and flowers bloomed. Fruit, nuts, and seeds invited the animals to dinner. Butterflies and birds flourished; pollination quickened. Color came. The brilliance of springtime arrived. The modern world began.

Leningrad, U.S.S.R.

Intourist sequesters me at the airport. Through a partition, as I sit alone, I can see hordes of Muscovites queuing for the same flight. Yesterday I dumped my extra tapes in a street waste dispenser near a construction site, quite hastily and with forced nonchalance, sure I would be discovered, arrested, and interned: I am here as a tourist, not a journalist. Without looking back I crossed the street and watched a tennis match for a while, keeping the waste dispenser in view, on the lookout for officials or zealous patriots. I realized I was being paranoid—who should mind that I am interviewing biochemists?—but I reflected on the unhinged Delmore Schwartz's quite rational perception, "Even a paranoid has enemies." At the airport I am equally on edge: two tape recorders in my overnight bag and more tapes than I can explain. But I am waved through customs to the plane ahead of the locals who, once I am seated, flood into the aisles with net bags and cardboard luggage. We are served ginger ale, given plastic bags for our fountain pens, and with inadequate cabin pressure we are delivered to the Leningrad airport.

There, to my surprise, Intourist has arranged for a representative to meet me, a slight, sandy-haired man with an excellent command of English but a bad stutter; he has worked at being casual. He invites me to try his pipe tobacco—he is partial to a burley. Would I be amused to learn the lighting fixtures at the airport were built by Westinghouse? He laughs. Leningrad is a network of rivers and canals—Russia's Venice. There are some 300 waterways here. As we ride down Moskovsky Prospect, a wide, sweeping, lovely avenue bordered by trees and majestic buildings, moldering gracefully, it seems what I would have imagined Paris to be a century ago. A third of it was bombed out by the Germans, but more than twice that has gone up since. "A lot of it is high-rise," the young man says with pride, "like New York City." At dinner in the hotel, a cavernous dining room that could easily seat 1500, a table of Americans sings "Happy Birthday." The view through the vast picture windows gives out onto the River Neva, mauve at dusk; the same view as from my room above. Before I leave to see Takhtadhzyan, uncomfortable over how little I know of what I am seeing here, and vague even as to what I am seeking, band music draws me back to the window. A naval detachment is passing by, headed for an ancient destroyer anchored at a dock across the way. I guess it to be the equivalent of an American military company, about 200 men. They are singing, very smartly in step with the music.

Arman Leonovitch Takhtadhzyan is the Director of the Komarov Institute in Leningrad, which is the principal botanic garden of the Soviet Union. He is a Soviet citizen and a member of the Party, but he is not a Russian. He is a Soviet Armenian. He has been in the forefront of the development of a general system of classification of flowering plants—of angiosperms—for the last several decades. He is described as a difficult, unpredictable, and exacting "boss." He is said to speak English well and to understand it when it is spoken to him although he loses some in a group conversation. He is said to have a broad understanding of the world, to be a man of catholic tastes. As with Oparin and other Soviet scientists, his political position within the U.S.S.R. is complicated; he knows his limits. During the long years when genetics as a concept was politically unacceptable to state policy, he had some problems. But the official view of such things eventually softened, and when this happened, his position improved. At a New Year's Eve party, which he did not attend, some of his colleagues presented impressions of him related to slides purported to be specimens of fauna which were in fact photographs of seminude women. This is cited as indicative of an informal and easygoing nature. Heavy-set with dark thick hair, he seems to blend with his office, a high-ceilinged, cluttered room, with dark wood paneling, that does not admit much light.

"There is no evidence about the origin of angiosperms unfortunately," he says. "No evidence at all."

"Chemical analysis?"

"No. It's just speculation because there is no evidence, no intermediate forms or intermediate groups, neither in modern flora nor fossil flora."

He does not seem particularly perturbed about this although he waits as though expecting a reaction from me. Having none yet to offer, I wait him out. "Angiosperms are very isolated and have many fundamental differences in anatomy, flower, and fertilization. Angiosperms have so-called double fertilization which is absent in other groups. That is very complicated and better I don't speak about it." He looks at me to see what I have to say about this. I wait. "I think that angiosperms originated as a result of neoteny. Neoteny means the origin of new organisms not from the adult forms but from juvenile forms. Some anthropologists think that man originated from early stages of monkey, of ape, and the same is true about many groups of animals. The same is also true, I believe, of angiosperms. Because in the morphology of angiosperms you see many infantile characteristics, many characteristics which are absent in adult gymnosperms but which you find in early phases of their development." Gymnosperms preceded angiosperms, I later discovered, by several hundreds of millions of years. They bear seeds but their seeds are unprotected from the outside world. Takhtadhzyan continues: "For example, carpels. The carpels are folded, that's why they are angiosperms. Seeds are protected in angiosperms by carpels. The part of the carpel which contains the seed is the ovary."

"This is characteristic of all angiosperms?"

"This carpel corresponds to the early stage of folded leaves because leaves at the early stage are folded like this"—he makes a cup of his hand—"and I think the ovary was very important in the evolution of angiosperms. Another factor is the coevolution of flowers with insects. Insects were pollinators, and under their influence many peculiar features of flowers originated. I don't know whether it is necessary to speak more about this . . . "

I say that I hope he will.

"The first pollinating insects were beetles. Even now the most primitive angiosperms are beetle-pollinated. Magnolias are pollinated by beetles. They were the first pollinators. Now there are many others, bees, butterflies, many and more specialized pollinators."

"How important are angiosperms to modern life?"

"Of course they are the most important of all for the human race. You can't imagine our existence without angiosperms. They provide us with food, clothes, shelter, medicine. The very origin of mammals —and man, of course, man is a mammal—is the result of angiosperms."

"How extensive are they, how widespread?"

"We have about three hundred thousand species of plants, maybe more. Two hundred and forty thousand are angiosperms. Almost all of our cultivated plants are angiosperms."

"Is it known how the first plants were brought under domestic cultivation?" I have moved wildly out of chronological context, but Takhtadhzyan replies politely:

"Do you want such a question?"

"Do you have an answer?"

"I shall think about it. Tea?"

His secretary enters at this opening, with tea and candy. We sip and munch and smile reassuringly at each other. Oparin's cryptic remarks are still fresh in my mind, and though I do have some fear for the question, I ask it anyway. "Which came first, plants or animals?"

"According to the modern view, they had a common origin."

"But one must have been earlier than the other."

"Blue-green algae are very, very, very old, of course, but blue-green algae are not algae from the modern point of view." He pauses; it is evident he is thinking about how to put this to me. "We have two groups of organisms, prokaryotes and eukaryotes—prokaryotes are cells that have no nucleus, eukaryotes are cells that have a nucleus. Blue-green algae and bacteria, the early forms, have no nucleus—no chromosomes."

"They are neither plants nor animals?"

"Yes. Both plants and animals originated from prokaryotes in ways we don't know yet. But there is an interesting theory that plants originated in the way of symbiogenesis. If this is true it means that plants originated from the symbiosis of animals and some prokaryotes —some blue-green algae. Chloroplasts, the green bodies in leaves, correspond to blue-green algae. They *are* blue-green algae. There was a symbiosis between heterotrophic animal cells and blue-green algae, and as a result—plant cells. If this is true, and some accept it and some do not, plants originated from animals, not the other way around."

This seems wondrous to me, and I say so. "Better not to ask the question," he says. "It complicates things."

Taking a paper from the pile on his desk, he now declares there is something more he would like to say. He will read to me. It is what he feels about certain misconceptions having to do with the importance of plants to all of life. He puts on his glasses and begins:

"Any ecological system includes in addition to its living components at least two main types of organisms: producers and consumers. They differ from each other on the basis of nutrition and the energy source used. Producers are mainly green plants, which are able to use light energy to manufacture their own food. Consumers cannot manufacture their own food, so they use organic molecules made by plants as a *source* of food.

"There are two main types of consumers—primary consumers, or herbivorous animals, which eat plants, and secondary consumers, or carnivores and parasites, which feed on the primary consumers. There are also the decomposers, or destroyers—mostly bacteria and fungi, which are responsible for the decay and recycling of organic material in the soil and water. About three hundred thousand species of green plants provide all the food for at least ten times as many consumers, including man himself. Because of the nature of food chains and the degree of specialization in feeding habits of these consumers, we may assume that the extinction of each species of plant is on the average accompanied at least by a tenfold loss among other organisms.

"There is constantly going on a selective interaction between the two major groups of organisms in an ecosystem—between producers and consumers. They have evolved together in the sense that the evolution of each is dependent upon the evolution of the other. The whole history of the organic world is the history of selective interaction of plants and animals. Probably the most important example of this coevolution is the role of pollinating insects in the origin and evolution of flowering plants—the angiosperms. It is widely accepted now that insects have played a prominent part not only in the evolution of the flower but also in its inception. Insects and flowering plants—angiosperms—have become the most numerous groups of organisms, and the flowering plants have reached the highest level of organization in the plant kingdom."

A living fossil, a relative of a most ancient plant, is the ginkgo tree, Takhtadhzyan says, returning to conversation. It is a gymnosperm, tracing its own derivation from a seed fern of some 200 million years ago. It is called the maidenhair tree, and the odor of the seed coat is said to be like rancid butter. "Where does it grow?" I asked him.

"In culture everywhere. But it is not wild. In 1958, when I visited China, a botanist there told me they have it growing wild in the east. But only in a small area. In New York you have some on your streets."

"Where?"

"On one of your avenues, I don't remember which. Ask Cronquist." Cronquist is Arthur Cronquist, a director of the New York Botanical Garden, an associate and friend of Takhtadhzyan. Ginkgos turn out to be found throughout the city; in Greenwich Village, several were planted by Mayor Jimmy Walker. "Cronquist has published a very interesting article," Takhtadhzyan says, " 'Adapt or Die.' It is on conservation." Later—its title seemed an odd imperative for the narrow world of conservationists, and I was all the more intrigued by the Russian endorsement—I tracked it down. It proved to be very much to the point of why I was here now, struggling to follow this Russian Armenian.

Cronquist delivered his paper in 1970 at a symposium in Belgium on the place of botanical gardens in nature conservation. It is a short, pointed, blackly pessimistic warning to the human species, which veers swiftly from the concerns of the flora to the "larger problem of human society in an equilibrium community," with little apology for the deviation. "Without such an equilibrium," he notes, "either the flora will be destroyed, or the society will be destroyed or will destroy itself, and the preservation of the flora will no longer be of any concern."

Thus, Armageddon from a botanist's point of view:

> We are learning that our air, as well as our soil and water, has limits to self-purification. A few years ago some of the big cities had unpleasant air, but now the poisonous stench often covers large regions. The smog which used to envelop Los Angeles has not only become more vicious, but it has spread to cover the whole valley. It is killing ponderosa pines eighty miles east of Los Angeles; and it is spilling over Cajon Pass, at an elevation of 1,300 meters, fifty miles east of Los Angeles, and spreading out into the desert. . . . Our civilization may be compared to a species, just as a biotic community has been compared to an organism. . . . As biologists we know the stringency of the evolutionary requirement of adaptation. . . . Our society, like a species, now faces a severe environmental stress, and it has the same alternative as a species: Adapt or die!

If, as Takhtadhzyan has declared in his prepared statement, the extinction of a single plant may order the doom of ten species of animals, the scope of Cronquist's concern comes somewhat clearer. The plant is more crucial to life than one might have suspected from an otherwise obvious truth: more crucial not only to human life, but to all animal life.

For the most part environmentalists tend to collect their concerns about the same obstacles to an orderly future: excess population (too many people for too little food); the exhaustion of natural resources (oil as the most immediately conspicuous example); pollution and the deleterious effect of chemical wastes; and nuclear proliferation. But for the botanist the reckless and swiftly accelerating destruction of the world's plant system, upon which all animal life depends, is a special horror. Takhtadhzyan turns to his concern for the world's rain forests.

The richest concentration of plant life, he says, is the jungle band extending around the world in the equatorial zone. By geopolitical coincidence the rain forests fall within the boundaries of some of the poorest nations on earth (Brazil, Zaire, Malaysia) for whom they offer a vital developmental resource in agriculture, cattle raising, and logging.

"About two-thirds of the higher plants occur in this zone," Takhtadhzyan says. From the botanist's view they are the principal source of plant production for animal consumers. Rain forests cover 3 to 4 percent of the earth's surface, but they produce from 20 to 35 percent of the world's oxygen and recycle roughly the same amount of its carbon dioxide. They have a significant but not yet fully understood effect on the weather. Damming up the Sudd in central Africa can change the monsoon over India.

For world animals the rain forests are a collective resource. Once gone, they are no more "renewable" than oil. Each nation has the sovereign right to do with its natural resources what it will. Costa Rica cuts its rain forest to raise beef cattle, which provide hamburger for American fast-food chains. Yet the separate destructions of this collective resource race toward a crisis for life that is little known and seldom discussed. To improve speed and efficiency in logging desired timber, an American industrialist has developed floating sawmills to cut their way up the Amazon. "In all," Takhtadhzyan says, "they are being felled at the rate of sixty thousand acres a day, about twenty-five million acres a year"—about fifty acres a minute, day and night. "Eighty percent of the modern world's rain forests have already disappeared." At the present rate, Professor Peter Raymond of the Missouri Botanical Gardens estimates that at least 50,000 species will have reached threatened status or become extinct by the end of the century. Unless the process is reversed—and there has been no collective effort to stop it, indeed little recognition that it is an existing problem—some ecologists predict that all the rain forests will be gone by the end of the century.

"How old are they?"

"There were already rain forests by about a hundred million years ago, maybe just a little later. They are dominated by angiosperms, the flowering plants. They are very complex organisms, the result of long evolution. You can take some parts away partially, but only some parts. You should preserve the main part."

"If the present rate of cutting them down should continue—"

"A catastrophe." It is the first time this polite man has cut me short. "It would be a real catastrophe."

5

WAITING IT OUT

Trout Lake, Washington

It was odd about the weather. Here it was Tuesday—Vihtelic made himself name the days, not just count them by the number he had been here, which now approached three—September 14, the middle of an early fall month fairly high up in the mountains of the Pacific Northwest, and it hadn't varied by more than ten degrees. Forty was the lowest, he guessed, which was far from unendurable, but the dampness was a pain in the ass—literally it wet his ass. Sunday and Monday had been overcast and Tuesday now was, too. Late in the afternoon of both days it had rained, and though he was sheltered for the most part by the car frame above him, the water trickled down through the driver's side and seeped into his sleeping bag. *He had to get out of here!* Where was the search party? Where were his brothers? He knew they would be looking for him. Lou would probably be out here beating the bushes right now, except that Lou was in Iran. Larry, Frank, and the others, then—when were they coming? Vihtelic rose to his knees, placed the wrench behind him under the dash and began to pry and twist. He had arrived at some threshold of pain with his foot, almost as though having made a bargain with it. So long as he didn't move, it didn't hurt him so much. Moving in any direction, but especially prying at it, cost him heavily. What else could he do? Being absolutely practical about it, if he didn't get out today he would lose it. He was working against a deadline and there was nothing he could do to change that fact. He wedged against the dashboard behind him until the pain forced him to stop.

But if they didn't come, what then? How long could he last? Vihtelic felt about his body, probing his fat supply—his stomach and inner thighs. If he could find a way to get all the water he needed, he was sure he could hold out for two weeks. By then, if no one had come, he would have to cut off his foot, using the sharpest of the glass fragments he could find. He knew a trapped animal could chew off its foot. In the army, as part of his medic training, he had amputated the leg of a dog. He knew the steps: you apply a tourniquet, you ligate the arteries, and then you cut through the bone. He knew he couldn't assess the pain this would bring, but if that was the choice, that was what he would do. He had to get out or he would be dead. If nobody came *he could die in here!* Vihtelic began to cry. But he *wouldn't* die in here, he would saw off his foot before he did. As quickly as he had started, he checked himself.

"I'm getting out of here today," he said out loud. "Don't worry." He forced himself through the beginning of the routine he had set for himself. He urinated, and then he combed his hair and wound his watch. He *had* to solve the water problem.

The solution probably lay somewhere among the resources available to him in the 1975 Mercury station wagon. He had no more cans or bottles. He looked again for the coffee pot—what he would give for that damn coffee pot! Rummaging on through the litter he found another T-shirt, and as he held it in his hand, turning and squeezing it, the thought came that the T-shirt might be the solution. He tied the shirt to his wire line, folded seven loops into the line and, resuming his chant, he tossed it hard out over the log and into the stream. When it sank to the bottom he began gingerly to pull it in. It slipped over the log, then over the sand and pebbles in between arriving covered with muck—but soaked. He sucked the dirty wet from the shirt. It worked! He threw it out again, slipping it up and over the rock, sucking out the water, throwing it out again. If he pulled too hard or jerked at it, the line would break. When this happened, Vihtelic said out loud to himself, "Don't worry. There's more where that came from." He pulled wire from under the dash, tied it to another shirt, and resumed his new routine. He threw out the shirt, coaxed it in and squeezed it dry until his belly was full, and then he brought in more to wash his face and the upper part of his body. Some of his anxiety abated: it was still too soon to tell, but it looked as though he might have solved his water problem.

By now it was the middle of the afternoon, when the warm breeze came. It arrived on schedule today as it had the last two days, and this time Vihtelic was waiting for it. It warmed him, and he slowed his routine to enjoy it. For the first time he noticed there were birds playing about the stream. Juncos probably, or some kind of sparrow. He whistled at them; then he looked out to the water across from him. Directly to his front a tiny little fountain of water leaped up into the air. It was forced up probably by a hole or a rock in the streambed. He thought how easy it would be, if he were free, to walk over and drink from it. It was beautiful and clear as glass, rising up out of the stream. He wished he could do that now. When finally they came that would be the first thing he would do.

He was getting used to the stream, used to its sounds, so sensitive to its frequencies that he could hear above it the motor of an approaching car from as far away, he guessed, as a quarter of a mile. He could tell by the sound whether it would be a camper, a car, or a truck. It surprised him that his hearing could adapt in such a fashion to a noise which he thought at first to be deafening. Probably it had something to do with his need, that he was straining to hear other sounds. In any case every moment of these cars driving into and yet above the confined world of John Vihtelic was known to him. They came out of the mountain to slow at the bridge, thump-*thump,* thump-*thump,* and then they picked up speed and drove on. If, for a second, he could see them, why, for that second, couldn't they see him?

There had been only three today and not one of them had seen him. There probably wouldn't be any more today. *Why* couldn't they see him? He had better get back to his water maintenance schedule. He had lost his foot, that was for sure, no matter what. The finality of his loss suddenly engulfed him. He screamed and yelled above the stream: "My foot! My foot! I'm going to lose it!" He began to cry, realizing as he did he had to stop this, that it did no good. It was give-up-itis. He made himself stop crying but it took him longer now than it had before. He had to go back to his water schedule. His throat was sore and the glands in his neck, when he touched them, seemed swollen. Exposed to the dampness, lying day and night directly above his body wastes, he realized he was vulnerable to a cold at least, maybe pneumonia. He shouldn't be surprised that his throat hurt. He returned to his water schedule, thinking of his family and of Mary.

The Cambrian

Nothing about the earth stays still very long. Weather shifts, seasons change, mountains form, continents collide and break apart—all of these things happening aboard the moving earth, speeding outward through space at 250 miles per second. Yet except for some instant cataclysm—an earthquake, an erupting volcano—its most visible moments of light and shadow, in ebb and flow, are so commonplace as to make it seem forever the same, as though freed of time. But in moving it changes; and animals and plants change, too, their destiny set by new circumstances. They change the earth in turn.

Levels of the earth are records of change, and fossils hold within them a record of the kingdom of animals moving along. Some living animals have survived more or less intact from the time when they first appeared (for the horseshoe crab, that was 500 million years ago). Through levels of time, solving old problems and creating some new ones, the kingdom of animals grew in shape, size, and ingenuity, and its boundaries spread. Even the interiors of modern animals show this architectural progress.

The sponge, that simplest of creatures, has a single tissue layer which encloses a central cavity. The jellyfish has two tissue layers, the inner layer enclosing the gut. The flatworm has three tissue layers, the innermost layer enclosing the gut. The roundworm has three tissue layers and a primitive body cavity. From roundworms to clams to octopi, through starfish, sea urchins, fish, and on up, the inner dwelling has become securely cordoned off: a fortification for the inner self as the cell protects its nucleus, the angiosperm its seed.

But it is the façades of animals, the architecture of their outsides, that make more quickly evident their response to past problems posed. Horseshoe crabs look much the way they did half a billion years ago because that way worked then and still does. Animals that don't solve problems die out or change. The problems are posed by scarcity, and by the changing world.

Among the first animals were burrowing worms, living within the floor of the sea. They developed an external skeleton with extended appendages, like oars, to propel themselves across the sea floor, thereby extending their feeding grounds. They were the first ancestors of the arthropods (which means "jointed feet")—crabs, shrimp, scorpions, and, later, insects. From the first there were floating animals, too, jellyfishlike creatures which caught food from the water passing through them. Some of the floaters developed a flexible dorsal cord and muscles to bend it from side to side. By 525 million years ago, some floaters had become swimmers, the first fish.

The earliest fish were armor-plated and jawless, ingesting suspended material in the water through open mouth and expelling the filtered water through gills. In time, anterior gill slits became jaws, and fins grew, in pairs. With paired fins providing mobility and balance and new jaws some choice of diet, fish branched out into a variety of forms, and two distinct lines arose: the ray-fin fish and the lobe-fin fish.

The ray-fin fish is the early ancestor of all modern fishes. The lobe-fin fish, which developed a bony structure within its fins, is the early ancestor of four-legged creatures and therefore of two-legged creatures as well. As it turned out, the lobe-fin fish (except for one of its kind, the lungfish) became extinct. This happens. Since life does not proceed in a straight line, animals come and go, their developed variations of proved value surviving them in other creatures of the same lineage.

Descendants of the lobe-fin fish came to the shore as amphibians, remaining close to the sea to keep their skins moist, returning to it to reproduce. (The development of a tadpole into a frog is, within the life cycle of a single living animal, an example of that epochal commutation.) To free themselves from dependence upon the sea, some of these ancestral amphibians produced an egg that could exist on land; they became reptiles. A group of reptiles, the therapsids, were distinguished by legs that lifted the body clear of the ground, and are believed to have given rise to the mammals.

From reptiles came dinosaurs, animals by their architecture alone often rising to the heroic scale. New hip structures extended their range. They lasted 150 million years, and then quite abruptly and without known cause, they disappeared. Whatever problems they previously solved they could solve no longer—either that or the problems changed, most likely the latter. In any case, seventy-five million years ago, as flowers bloomed and dinosaurs died, animals of all sorts—whatever their lineage, so long as they weighed more than twenty pounds—were wiped out.

Prominent among the survivors were the small arthropods, the first animals which, among all the members of the kingdom of animals yet to come, remained versatile and persistent. They were the first to fly. One of their kind, the scorpion, was the first air-breather to move onto land. As insects they exist today within any environment capable of supporting life. In England there are as many species of beetles as there are mammals in all the world. If the estimate is correct that the total number of species of living things on earth numbers between two and three million, more than half of them are arthropods—insects—still flourishing.

Fort Collins, Colorado

Some rare people write as they talk; there is no barrier between the world they confront outside themselves and the familiar furniture inside their heads. They say what they have to say, naturally and with ingenuous ease. In such people the voice is inherently unique, comfortable, appropriate to the occasion. Some writers work to talk as they write (Hemingway, for example) but the disparity is soon obvious, weakening the force of their presence. Still others—most others—make no effort to reconcile the two, the voice of their inner world so private a correspondence as to seem the work of someone else, stranger to the owner.

Late in the day, at the zoology building of the University of Colorado, I got on the elevator with a tight-lipped man of country bones and mud-red hair, wearing dirty work clothes and black canvas shoes. He had a box of cans and rags beside him, and I took him to be a maintenance man. I spoke to him, and he said, "Can I help you?" I said I was looking for Dr. Evans. "I'm Evans," he said. I noticed now that his equipment included a butterfly net.

Howard Ensign Evans is a farm boy from Connecticut whose first job was picking hornworms off tobacco for a penny each. Insects have always beguiled him. He is now an entomologist; a former curator of the Museum of Comparative Zoology at Harvard; a specialist in the study of wasps and bees, and author of, among other works, *Life on a Little-Known Planet.* From its first page, this book opens into a most private correspondence between the author and his inner self:

> . . . a summer's day on my back porch, as good a place to contemplate the universe as any, and better than some. The calling of blue jays and the rustle of leaves remind me from time to time of the richness of this our

earth, and the drone of traffic on the nearby superhighway, accented now and then by the cries of children from neighboring houses, remind me of man's dominion over it. I warm my bones in the sun, mother sun, without whose energy no leaf would rustle, no traffic drone. The summer sun evokes a dreaminess unlike that of night; it is as though one's flesh and bones had deliquesced, had somehow slipped back for a moment into the elemental stuff from which they were long ago raised and sent upon their strange career. I seem to see a world of searing flame, of forms without form, of shrieking silence. Then a dog barks, and I know I have been asleep.

A fly pivots on the rail. Does he know he is trespassing on the sanctum sanctorum of man, the suburban home? Why does he appear so self-assured, so unmindful of me? Doesn't he know he was once a mere maggot, bred in a garbage pail? . . . Evidently he concerns himself with none of these things, nor even with the fact that I, owner of this porch, represent the ultimate form of life on earth, the apogee of evolution, the image of God. Of course, after forty-nine years I do not have nearly so many offspring as he has fathered in a few short days, and I most certainly can't make a forward somersault and land on the ceiling. But I know all about him (well, quite a bit anyway), and he knows nothing about me. And my kind has learned to split the atom and to send vehicles into space. Knowledge is power; and we could reduce the earth to rubble if we wished (and even if we didn't wish). What has the fly done? (Come to think of it, this may not be a very fruitful line of thought. I have read somewhere that insects are several times as resistant to radiation as man, and I recall how flies revel in rubble and corpses.) I flick my hand, and the fly is gone. . . .

Inside his office, spare and windowless, Evans does not now seem to be the sort of man given to such aimless conceits. He will answer what questions he can but he doesn't know everything. Also, he wants to be away from here in an hour or so. There is a meeting tonight over the proposed reclamation of a wildlife reserve for mineral exploration, and he intends to be there.

What I want to talk to him about is simple enough to ask but complicated for him (or anyone else) to answer. The eminent zoologist G. E. Hutchinson asked the question himself in the title of an oft-cited essay, "Why Are There So Many Kinds of Animals?" But Hutchinson's answer was, for me, hopelessly embedded in the unyielding world of academic terminology. ("We may, therefore, conclude that the reason why there are so many species of animals is at least partly because a complex trophic organization of a community is more stable than a single one, but that limits are set by the tendency of food chains to shorten or become blurred," etc.)

I hoped Evans might bring the question up onto his back porch, might even take it back a step to explain as well the *progression* (if that was what it was) of animals of one kind to the next. If at first (to use the same simple directness as Hutchinson) there were only sponges and now there are elephants, how did the one lead to the other? A man who daydreamed about houseflies ought to be able to tell in a somewhat more general way how such things came to be. Or so I hoped.

Tentatively I begin to probe certain aspects having to do with the past several hundred million years.

Is it possible now to date accurately the major developments in the emergence of life?

"Yes, we're pretty good at that. We find fossil evidence in ancient rocks, and the age of the rocks can be established by radioactive degradation. More recently of course we have used carbon. You'd have to talk to a physicist to find out more about that."

Unpromising. I try another tack.

Where do the words come from for setting the geologic names of the ages? The spans of tens and sometimes hundreds of millions of years?

"From the locality where a certain stratum was found. Pennsylvanian, for example—this particular stratum was discovered in the Pennsylvania coal fields."

What about Cambrian?

"I'm sure there must be some place called Cambria but I don't know where it is."*

I seek to lead him. Wasn't the Cambrian period of 500 million years ago when the diversity of life began?

"Well, there's a big mystery there, because in the Cambrian all of a sudden there *are* a great many different kinds of animals in the sea —all of them in the sea, of course. I'm not sure people really understand why this happened. One theory is that calcium carbonate became available about this time to marine organisms. They used it to build shells."

Is there an explanation for this?

"There may be but I don't know one."

If the animals were all in the sea, was the earth covered by water then?

"You're getting out of the area of my expertise. My impression is

*Wales, I later discovered.

that the continents had formed but were uninhabited. The continents were much different from what they are now. In fact at that time there was almost certainly only one continent, all squished together in one big land mass."

Are there other theories about that time?

"Certainly, but I'd have to look them up."

He is giving nothing away. I realize I must have miscalculated the temperament of the man. Impossible now to imagine him daydreaming much about anything. No other course evident, I press on: What is the ratio of extinctions to survivors among today's life forms?

"You can safely say that through geologic time most organisms are extinct. Let's say there are three million species now living. Over time there must have been a hundred million species which are extinct—a rough guess."

"The implication being that life is all the more precarious?"

"Yes. Extinction is the rule. Most organisms either (a) become extinct, or (b) evolve into something else."

But the horseshoe crab, which appeared 500 million years ago, is still around? He studies my question and the textbook picture I have placed before him.

"Horseshoe crabs have persisted. They go back to the Cambrian, and they are quite similar to present-day ones. They have evolved, however. The paleontologist may call this one a Cambrian species, another an Ordovician species, and so on. But it's actually one evolving stock. True extinctions are groups in which all the members become extinct." Still, he says, lingering over the picture, the horseshoe crab of the present does look strikingly similar to fossil casts of horseshoe crabs which drifted the earth's seas half a billion years ago. He glances at his watch.

A half billion years! Evans's equanimity is not much disturbed by such a marvel, perhaps because he has been aware of it for so much longer a time than I. As we look together at the printed comparisons of the ancient species to the modern, I find myself reflecting on the ancestry of us observers. Within the kingdom of Animals, the phylum of the Chordata, class of Mammals, order of Primates, and the genus of *Homo,* we are sole surviving species. Against the 500-million-year-old horseshoe crab, we are able to date our own arrival on earth back to not much more than a million years! At late afternoon, in Fort Collins, Colorado, however (and with Evans growing a trifle impatient), it is we who are studying the horseshoe crab and not the other way around. It seems perfectly reasonable now, therefore, to ask, and I do:

"Is the basic point of evolution the tendency of all organisms to move toward complexity?"

He looks at me blankly, as though I had slept through the semester. "The basic direction of evolution is toward adaptiveness to the environment at that particular time. No. It's *not* toward complexity."

By now, the informed reader will have surmised from Evans's part of our conversation that the answers sought by me for larger questions as yet unasked were supplied largely by Charles Darwin early in the second half of the nineteenth century. Such a reader who chooses to get on with the discussion, as Evans is so clearly eager to do, may turn ahead to page 62, where our conversation is resumed. However, for those like myself whose familiarity extends not much beyond the catch phrase "survival of the fittest," Darwin's theory of evolution by natural selection, which now is seen to be the organizing principle of modern biology, deserves more careful attention. Indeed, in fairness to Evans, what he says from 62 on will make little sense without it.

There is a nice symmetry in beginning with Charles's grandfather, Erasmus Darwin, a medical doctor and the author of an epic poem described by a contemporary as the best bad poem in English literature. Poetical speculation in eighteenth-century England was a more prudent way of dealing with controversial ideas than stating them outright, and Erasmus Darwin used his poem to put forward his own version of natural history. Plants and animals descended from a common ancestor; life has since evolved as plants and animals adapted to a changing environment, more of them dying in the process than surviving. The abundance and diversity of life were the consequences of this evolutionary sequence.

Among those who pondered Erasmus Darwin's heretical speculations and were influenced by them was the Reverend Thomas Malthus. Eleven years before Charles Darwin was born, Malthus wrote an essay on the inherent limitations of human populations, suggesting that the numbers of people are governed by their food supply—in essence that there are always more hungry mouths than food to feed them. Populations increased exponentially (that is, in a geometrically increasing ratio), Malthus said, whereas food increased arithmetically, and this had always been so. Competition for food must be seen as a biological fact of life.

One now comes to Charles Darwin himself, a dropout candidate for the clergy and an amateur student of natural history who, in 1831, joined a mapping expedition into the southern hemisphere, with the assignment of collecting and observing the plants and animals of primitive places. Arriving at the Galápagos Islands four years after his departure from England, he wrote in his journal: "Here, both in space and time, we seem to be brought somewhere near to that great fact—the mystery of mysteries—the first appearance of new beings on earth."

Otherwise, what more he might be seeing at the time he was not yet exactly certain. Returning home, he studied and reflected on the evidence he had collected and the sights he had seen. Among the mysteries were some birds on the Galápagos, specifically finches—"Darwin's finches," as they have since come to be known (although at the time it took a taxonomist at the British Museum to tell him that was what they were). The finches were present throughout the area, but they were not all the same. There were thirteen islands in the archipelago—and thirteen different kinds of finches. Some had heavy bills for cracking seeds; others long bills for taking nectar from flowers; still others had developed an ability to seize cactus spines in their bills to probe for insects. How could this be? The archipelago was isolated—a distance of 600 miles from the mainland. Had each variety been created to live according to the needs of its respective setting? He wrote, "One might really fancy that from an original paucity of birds in this archipelago one species had been taken and modified for different ends." Back in England, he worried the question for several years.

One day, to get his mind off his own work, Charles Darwin turned to Malthus's essay (as Malthus had turned to Erasmus Darwin's poem). Contemplating the inevitable disparity between consumers and producers, he recognized an essential condition that led directly to the theory of evolution through natural selection. Darwin reasoned that not only man but any animal had to compete for the conditions of its own survival—for food, light, water, or whatever it took to live. Competition was life's spur, and competition was endless—against one's own kind, against other kinds, against the indifferent, ever-changing earth. Losers produced fewer offspring than winners. Over hundreds of millions of years, the altered earth was evidence of changes ordered, and fossils of animals now extinct were surviving, fragmentary evidence of those times, teasing clues to their past existence. The present was the most recent record of new and ingenious solutions to problems of the past. Interwoven into an endlessly complex design for mutual survival, plants and animals living now were heirs to the world estate.

But if Darwin could believe this grand scheme was correct in the long run, how did it work in detail? What caused the finches of Galápagos to become "so many kinds"? And if all had earlier come from a stock somewhere else—on the mainland presumably—what was there before finches? Before birds? In a single square foot of soil there may now be found the seeds of dozens of different kinds of plants and hundreds of different kinds of animals, diversity of a bewildering order. Even allowing for so extravagant a perspective of time and space, the living evidence of surviving solutions seemed so vast as to overwhelm any attempt at a comprehensive explanation.

To follow through the implications of Darwin's answer, consider further the example of the finches. Parent finches tend to produce more offspring than the environment can support. No species of finch is exactly like any other; in fact no single finch (no single anything) is exactly like any other. Among the varieties of individual finches there exist certain characteristics which tend to favor one over another in their competitive struggle to survive. These variations may be so slight as to be undetectable until, over time—through generations of populations of offspring—they emerge as characteristics of discernible value, as they had on the Galápagos when Darwin noted the finch with a bill better suited to cracking seeds, or to capturing insects. Through natural selection—nature being the selector—the seed-cracking finch had adapted to its environment. When such finches mated between themselves to the exclusion of all other finches, a new species had formed. For the moment, within that quite specific environment, the seed-cracking finch had, through its new identity, won life.

As with finches, so with all else, from the lowest of organisms to the highest. There was a time when DDT introduced into the environment of houseflies killed them off en masse. Some few survived, however, and in reproducing themselves, the survivors eventually produced new generations of flies resistant to DDT—natural selection working its ingenious unstoppable ways. Understanding the processes of natural selection helps explain phenomena on a larger scale. The annual use of insecticides has increased in the United States over the last forty years by tenfold, to about a billion pounds. Yet because some 400 insects and mites have developed resistance to those insecticides, crop losses caused by pests have increased from 7 to 13 percent. By studying domestic breeding techniques, the similarities and differences of wild plants and animals he had observed, the comparison and analysis of physiological features—by all such methods did Charles Darwin systematically assemble his evidence and draw his conclusions on space, time, and struggle in the production of life. The wing of the bat, the flipper of the whale, the arm of the man are appendages of common origin, traced to primitive reptiles. The bones in the human face lead backward to the gill arches of ancient fish. The intricately complex mosaic of present-day life is the sometimes direct but most often indirect consequence of all that has gone before.

Still, Darwin in his time remained uncertain as to the nature of the exact characteristics within any creature which provided for such life-favoring change. Certain variables were present in each individual and they were inheritable, he was sure, but how nature selected from among the many variables to sustain life was another mystery, one made all the more vexing since natural forces were hardly constant. Nor was it clear how direct the relationship was between cause and effect. If over eons a river carves a valley out of a mountain and all the animal life thereabouts is affected, has the river caused the change? The answer—partly yes, partly no—is at the heart of an elegant subtlety to Darwin's theory of evolution by natural selection, and although it was there within the theory he propounded in the mid-nineteenth century, Darwin would not be able to work out its particulars in his lifetime.

Six years after Darwin published *On the Origin of Species by Means of Natural Selection,* an Austrian monk (another amateur), working with garden peas to find out why some of them were wrinkled while others were not, arrived at a beginning to *that* answer. Inheritance within an animal or plant—the variable, the *mechanism* of change—is controlled by genes, which are fixed within chromosomes, which reside within the nucleus of the cell.

"Well, then," I said to Evans, hoping to close quickly the awkward gap, "are there examples of life moving toward greater simplicity?"

"Parasites invariably become more simple in structure. Complexity and simplicity are pretty big words. Every animal has certain complex features and certain simple features. Houseflies have some complex structures and some simple ones. The struts on their wings, for example, are extremely simple, and in that respect they are simple organisms." (But in other respects, as Evans pointed out in his book, they are complex, too: "The middle segment of the thorax, which bears the first and only pair of functional wings, is greatly developed at the expense of the other two segments, forming a robust box crammed with powerful flight muscles. These muscles are of the indirect type; that is, they move the wings by changing the shape of the thorax itself.") "We have very complex brains," Evans goes on, "but we too are simple in some respects. Our teeth, for example. Our dentition is simpler than that of many other animals."

"What is a measure of success in evolutionary terms, then?" I asked. "Just survival?"

"Well, sure. But success is *also* a hard thing to define."

"Wouldn't insects as a class be more successful than most other forms of life?"

"That could be argued, yes. In terms of numbers of species insects exceed even the plants. Let's say there are a million species of insects—a fair guess. The number of flowering plants is much fewer than that, even if you take all the plants, including ferns and so on. Are insects more successful than we are? We are only one species. Our criterion of success is how we thrive at the expense of the environment. That's not the way insects would define it. A million species in the world, all of them perfectly successful. Yet they don't dominate the environment."

They have come close, however, in one way, which could be seen as yet another measure of their success. Evans believes they account for the downfall of the dinosaurs, a calamity which thus far has been explained to no one's complete satisfaction, though Evans's version satisfies him. "Butterflies and moths evolved in the Cretaceous"—the age between seventy and 135 million years ago. "Butterflies and moths have caterpillars as larvae, you know, and caterpillars, such as that of the gypsy moth, can be terribly devastating. My theory is that when the flowering plants evolved, the butterflies and moths underwent an abrupt radiation. They ate up so much of the vegetation nothing was left for the dinosaurs. A bit later wasps evolved, and they are of course parasites of butterflies and moths. The wasps caught up with the butterflies and this gave the vegetation a chance to flourish again. And *this* shift gave mammals a chance to evolve in the Tertiary." We exist, according to Takhtadhzyan, because of angiosperms; according to Evans, because of butterflies.

"How long did all this take?" I asked.

"Oh, about a hundred million years." Now into his specialty, he is letting out a bit—not yet to the voluble, down-home garrulity of his writer's voice, but he is beginning to come around. At least we are in the same room.

"My understanding," I said, heading back for the links between the horseshoe crab and the elephant, "is that Darwin's theory was dominant through 1925, and then for the next twenty-five years or so, because of findings in genetics it fell into disfavor. What brought Darwinism back?"

A long silence and another baleful look. "Your dates are a little off. Mendel's laws were formed in Darwin's time but they weren't discovered until about 1900."

Evans resumes on page 65.

Although Charles Darwin couldn't see exactly what the components were regulating inheritance, he suspected they worked in an indirect fashion. The Austrian monk, Gregor Mendel, experimenting with garden peas, couldn't see them either (nor was it Mendel who named the components "genes"; that came later). But by breeding and cross-breeding generations of plants, Mendel was able to demonstrate *how* these components worked. Contrary to the traditional belief of plant and livestock breeders, an offspring did not simply reflect an "averaging out" of the characteristics of both its parents: that is, crossing a tall horse with a short one would not inevitably produce a middle-sized one. Rather, these characteristics of change were separate and distinct units in each of the parents, some dominant and some recessive, and they appeared in varying ratio through succeeding generations of offspring.

Early in the twentieth century a graduate student at Columbia named Walter Sutton established that Mendel's units of inheritance —the genes—were fixed along the chromosomes within the nucleus of the cell. The number of chromosomes in a cell varies by species. A potato cell, for example, has forty-eight. A human cell has forty-six.

In the human cell, twenty-three chromosomes are received from the mother and twenty-three from the father. Metaphorically, the chromosome is described as the vehicle for the message of inheritance, which is the gene, itself composed of deoxyribonucleic acid, or DNA. It was the accomplishment of James Watson and Francis Crick in 1953 to show *how* the message was delivered.

In any species the sexual process greatly increases the possibilities of variance of individuals. Within an interbreeding population of organisms, according to P. B. and J. S. Medawar, a single individual's genetic structure is only one of possible combinations numbering 10^{3000}, a variety of nearly unimaginable magnitude.

But the genes compose only the *potential* of an organism. How the organism actually turns out rests on the interaction of its genes with the environment—"nature versus nurture," in yet another shorthand reference to a fundamental biological tension. Genes provide the occasions for diversity and change—in the short run, the differences between individuals of the same species; in the long run, the inexhaustible possibilities for new forms. A key to it is the slowness of the process, relatively: favorable mutations persist, unfavorable ones are eliminated. A mutation in an individual organism is a static thing. It is only populations that can become increasingly better adapted over time.

In the early part of this century, as Evans has just said, with the rise of Mendelian genetics Darwinism faded in its force. Subsequent research, however, showed that Darwin's principles of natural selection worked not only on the visible level—as with his finches of the Galápagos, whose variations could be related to external circumstance—but on the cellular level as well, within the inner workings of an organism. It became clear, moreover, that these principles were true for all of life, at just about every level. Evolution is seen now to be chiefly the natural selection of genetic differences—fusing the two great perceptions of Charles Darwin, an amateur naturalist, and of Gregor Mendel, who couldn't qualify to teach high school science. Both men had inferred cause from the effects they had witnessed among animals like those nondescript finches and a plant as ordinary as the garden pea.

"The period in which Darwin was in the shadows would have been from about 1905 to 1927, I think," Evans says. "That was the time when some things were published pointing out there was no conflict, that genetics really supplemented Darwin's ideas. And a little bit later Julian Huxley came along with a whole series of books, *The New Systematics,* and there were others, too, and all of them pointed out that genetics was just what Darwin needed."

"The sources of variation were on a more minute level?"

"Sure. Darwin didn't understand this. He believed in blending inheritance—he had no idea that there were units of inheritance, which Mendel later discovered."

" 'Blending inheritance'?"

"He didn't really understand the process. He just called it that—a fusion of characteristics of two different organisms. But of course Mendel showed that it was particulate and others went on to elaborate on that. Nowadays we have all kinds of techniques for analyzing the genetic variations in populations to see the origin of species through geographic variation—that sort of thing."

Darwin was just as much interested as Hutchinson in why there were so many different kinds of animals. Moreover, throughout his trip about the southern hemisphere, he observed with growing interest that the same kinds of animals were, in a manner of speaking, different in different places. While this was true from continent to continent, and even within different regions on a single continent, it was most dramatically evident within neighboring islands. Islands were floating time capsules; hence the great significance to Darwin of the Galápagos.

"We think most speciation occurred in geographic isolation," Evans goes on. "Populations have to be isolated from other populations to evolve species characteristics. Islands—archipelagoes in particular—are excellent places to see this. Let's say a population—or just a mated female, for that matter—gets from Ecuador to Galápagos and is able to survive there. She is in a totally different environment and completely isolated from Ecuador. No others arrive. Now her offspring can evolve with no exchange of genes at all with the parent population, and they are going to evolve rather rapidly into a new species. If this species is in an archipelago, it can over time populate the entire archipelago, and these in turn will be isolated from the parent population."

"To the point of being unable to reproduce within that same group of islands?"

"Right."

It is growing late, but Evans does not seem to notice. He has at last become as much interested as I am in what he is saying. "It takes a little time, of course, to develop the isolating mechanism. But suppose one species got into an archipelago with five islands, and in the course of twenty thousand years developed into five species. Then you've got the potential of each of these species invading one or more of the other islands. So individual ancestral species might proliferate into, oh, a hundred species, I guess. And this has actually happened. In the Hawaiian Islands you've got an archipelago of five or six major islands, and there have been inoculations of insects from the Australian region over time—probably during monsoons—carried by winds, presumably. In the Hawaiian Islands you've got certain genera which have just gone hog wild. In one of the genera of wasps that I work on there's well over a hundred species."

"Over how long a time?"

"Not more than five million years certainly. I don't think the Hawaiian Islands are any older than that. There are other cases of very rapid speciation there—in a few tens of thousands of years. You've got groups like the honeycreeper, a group of birds which undoubtedly evolved from a single ancestor, and yet evolved into a whole bunch of species. But these species are very narrowly adapted —to a very distinctive fauna and flora and climate and so on. They're not able to compete with the continental species."

"Why is this?"

"Island faunas are very fragile, because the species are narrowly adapted, whereas continental species have a much broader genetic basis, more genetic variation. They can occupy a variety of habitats. Island after island has been defaunated and deflorated by man's introduction of things—continental species such as ants, birds, goats. You could make a long list of islands that have been completely ruined simply by the introduction of foreign plants and animals. Every time any species becomes extinct—and God knows a lot of the Hawaiian species are gone, many of the insects, most of the birds—we lose species which might be useful to us, at least as genetic potential."

"Is this true of plants as well?"

"Sure. As we grow plants we have to keep developing new varieties all the time, varieties which are adapted to different parts of the globe and resistant to certain diseases there, to insects, and so on. To develop new grains—new types of wheat and corn—you have to introduce new genes into the stock. Where are you going to get these if the ancestral plants are gone? There are lots of species of plants which are potentially useful to us, but we may not know it at the time. A good example is the jojoba tree which grows in the West. It was considered a weed. Now it's found to contain a rubberlike substance which can be used to make tires! There are lots of plants like that. We don't know their possible use right now but some day they might be just what we need."

Like the others before him, he is unwilling to view his science as separate from broader concerns. Carrying a reader across his little-known planet, Evans in his private writer's voice conducts a wondrous excursion. He examines the aerodynamics of the dragonfly ("The wings beat once in response to each nerve impulse, at a rate of up to about thirty beats per second"), celebrates the cricket's song ("What was it Nathaniel Hawthorne said of the tree cricket? 'If moonlight could be heard it would sound like that' "), and zealously as a sleuth, he takes an entire chapter to unravel the mystery of the firefly's glow. But he ends it all as a depressed scold, with Arthur Cronquist, worried sick over the thoughtless damage done by his own species to the antique earth.

The world has become urbanized; especially is this so in his own country. Three million acres of land are leveled each year; in California, bulldozers turn 375 acres a day. "Supposing," he writes, in reflecting on the consequences of pollution, "we inadvertently contaminated all our soils with substances that prevented the growth of nitrogen-fixing bacteria and the bacteria of decay. It would take no more than that to render the earth sterile of all life, including our own."

Yet against such fears he does not categorically advance the world of nature over his own. DDT stopped a typhus epidemic in Naples during World War II, he notes, and it has improved the health of people across the world through malaria control, saving five million lives and preventing 100 million illnesses. "These are a good deal more impressive figures than those on the number of deaths from the misuse of pesticides; about a hundred and fifty per year in the United States. . . . On the agricultural scene the bold fact is that we could not feed ourselves and the rest of the world adequately without the use of insecticides."

It is now quite noticeably late, near the end of our second hour.

From his book and from what I have got through conversation it is difficult to know on balance how Evans feels about the immediate future. Deliberately and still leading—even in his enthusiasms the man does not seem disposed to volunteer anything—I try to move him toward a conclusion.

"It's said that man is the first creature to gain control over evolution. Do you think there will be a price to pay for this?"

"What do you mean?"

"Will he have to pay a price for his independence from the laws of natural selection?"

"I see no reason why he should pay a price for independence from the laws of natural selection. No. He's embarked on a program of cultural evolution"—evolution of the mind through the accumulation of knowledge and culture. "The price he is going to have to pay, of course, is based on the fact that he's still ultimately dependent on the environment. He has to have a place to live. He has to have oxygen, plants, certain animals at least, and he can't really foresee his own future. No, his future is going to be cultural evolution, not a biological evolution, and we can't really see where it's leading right now."

"But with cultural evolution, doesn't he run the risk of leaving behind his biological origins—his awareness at least of these origins?"

"Oh, he can't afford to leave them behind. What I mean is that he's not going to evolve any more biologically—at least not appreciably. But he is going to evolve culturally. He can't leave behind his biological heritage. Man is an animal. There's nothing he can do about that."

There comes to mind the latest response to OPEC oil increases, the subsequent shortages, the subsequent hysteria, the subsequent relaxation into complacency even though the oil—that residue of compressed plants and marine animals of hundreds of millions of years ago—is known to be running out. "But isn't it possible that if you have cultural evolution to the point that the average urban family in this country believes that its natural state of existence is to have water, say, and some amount of fuel energy available to it, that cultural evolution has already progressed to the point where that family's relationship to its biological origins has been—"

"Well, if we reach the stage where the majority of people take that attitude, of course, then the human species is on the verge of extinction."

I wait for some qualification to this jolting pronouncement from one who has conspicuously backed off from more conservative speculations. None comes. The assertion hangs in the air like a sharp, unexpected slap. I wonder if I have overstated my question. Few people I know really believe there is a limit to the quality of the life they lead. Perhaps what he means is more applicable to Third World countries. I suggest as an example: in the Middle East the trees are gone, and charcoal needed for cooking is premium, imported from East Africa where scrub brush is cut and burned. For the people who need it—Asians existing on one hot meal a day—charcoal is a vital commodity. "Is it in this sense that you mean—"

"Essentially those people are on the verge of extinction," he says. "Once they can no longer obtain what they need—which will happen, I suppose—they will be eliminated. This is the crux of the whole matter. Without an awareness of the environment, man is doomed." He means all of us.

"Is this as a result of our failure to understand biological kindredship?"

"You can see it very well in the meeting I'm going to tonight. The purpose is to defend a wilderness area near here and to try to expand it. A lot of it is underlaid by coal and other minerals. What do you do about this? It's a serious problem. People have to have coal, and they're going to have to have a lot more of it as soon as the oil runs out."

"What kind of argument will you make?"

"I don't know. It's a very difficult thing. It seems to me ultimately we're going to have to rely on renewable resources. Coal is nonrenewable, like oil, of course, and even though we turn to coal and get every goddamn bit of it out of the ground, still the coal's going to end someday too."

Evans sees as a consequence of these diminishing resources and the inversely swelling proportions of people the ultimate problem for survival. He narrows to a surprising target: "A lot of people figure the hell with the earth. We'll move somewhere else in the solar system. This is absolute garbage. We're never going to have any place but the earth. A lot of people aren't aware of this—they gobble up all this stuff about the space program! I've just been amazed at people like Carl Sagan, whom I admire in some ways, but he's always talking about life in space, going to other planets. And now he's talking about going to a moon of Saturn, and by God he was hard to persuade there's no life on Mars." Evans's voice has not risen, his inflections as quiet and restrained as in his account of the butterfly–dinosaur cycles. "These monitors trying to detect messages from space, or to send messages out! An utter waste of time and money. If anything evolved on another planet it would be so utterly different from us that we couldn't possibly communicate with it. It might not even be life as we regard it."

"The island concept removed by planetary time and space?"—here is the other side of Ponnamperuma's confident assertion.

"Sure. We know pretty well how life evolved on earth, and we know how the first molecules came together, more or less. It must have been utterly different on other planets. Different chemicals may have been involved. The whole sequence of events on earth is so unique it *couldn't* have occurred anywhere else. It may be true there's something on other planets in other solar systems, something alive in the broad sense of the word. But the idea of man moving anywhere else—even to a space station—is an idea that repels me. The idea of creating a station in space! These stations are going to depend entirely on the earth for their resources. So what the hell good are they? We've already drained the earth. Now we want to build a space station depending entirely on the earth for its resources. It's just stupid. And the people behind these ideas call themselves *scientists!*"

It is after eight now and he must go. He is late for his wildlife meeting.

6

SETTLING IN

Trout Lake, Washington

Of the nine Vihtelic children, he and his sister Martha had probably had the hardest time finding a place for themselves, getting into something they would enjoy. Martha was as big as he was, and that was pretty big, but she was beautiful, too, and her size and weight were proportioned attractively. After graduating from Northern Michigan, she'd tried a lot of things, no one of which seemed to suit her. She would be okay, she was still looking. Until Drake Willock that had been Vihtelic's experience, too. He liked California and the laid-back style out there, and he would have stayed on if he had found a job. But he hadn't and he'd come back, working briefly as an X-ray technician before enlisting to beat the draft. Then he had gone to one army school after another—electronics, paratroopers, Green Berets, medics—anything to stay out of Vietnam; and when he got out he had a year at Hope College, in Michigan. If he'd gone on, he probably would have become a doctor, but he and Mary were going together by then, so he had taken a job at St. Mary's as a technician. Then came Drake Willock, and suddenly, without having known exactly where he was headed or why, Vihtelic had a future. A field representative for a big company manufacturing kidney dialysis machines.

A week ago last Sunday, to celebrate, he had thrown a kegger for himself and his friends; he had a lot of friends. There must have been eighty people there. One hundred and fifty ham sandwiches, thirty gallons of beer, all to wish Vihtelic good luck in a new life. Hello Drake Willock, goodbye Drake Willock. All the company knew about him were his references, and that wasn't much. Probably now, by Wednesday, Drake Willock thought he had just split with their car.

He put the heavy plastic shield he'd found under the dash against the frame beneath him and began to hammer at it with the wrench. If he could just crack a sharp edge out of it, then he could try sawing the log root behind him. He hammered and cracked but the edges were blunt. No go. Well, as long as he was taking a break from water maintenance he might as well try to hammer out some of the bumps in the roof to make his resting place more comfortable. He banged awhile on the bumps without making much difference there either, and then he turned his mind to other problems.

Having accepted the fact that he had lost his foot—that no matter what else happened, that part was over, his foot was a goner—he was able to relax some of the pressure on himself. To this extent at least, his position had become clear. He would continue to work at freeing his foot but the urgency for doing so had passed. The most likely thing to happen next would be for his rescuers to come and get him out. He had better rearrange his priorities now with that in view.

The first thing was to maintain himself properly until they came—stay filled with water, get as clean as he could make himself, stay warm and as comfortable as possible, sleep as much as he could, and conserve his energy. This part alone was something he would have to work at. By now he had found that in an average hour he would sleep and wake five times, and haul in water ten times.

The second thing was to let people know he was down here. Nobody said he had to wait until the search party found him. He simply could not comprehend *why* the cars passing above hadn't seen him; but he was convinced that if he worked at the problem methodically and with patience he could find some way to make sure they did.

So—back to inventory. He pulled his rugby shirt out of the pile in the rear of the wagon. Its blue-and-bright-red stripes were like a signal flag. He draped it over the edge of the chassis above him. As he twisted about, the vanity mirror on the sun visor caught and gave back his dark visage. For the first time since he had been in the hole the sun was coming out, edging along the tops of the trees, driving back the shadows of the dark ravine. The mirror! Maybe he could bounce the sun's reflection from it up along the road and catch a driver's eye. But reaching out with it, he found that he couldn't extend the mirror far enough beyond the edge of the window frame. He looked through his stockpile. He could use the tennis racket. He fixed the mirror onto the face of the racket with strips of masking tape. As he worked, the sun warmed the car, then warmed him inside it. A welcome change, but then a sharp acrid smell began to rise up from beneath him, the smell of gasoline. Apparently its fumes had been held down by the dampness of the past four days but now it was all over the place. *Jesus, what if it should ignite?* The prospect of it brought a new fear which he respected—it scared the shit out of him—but could not allow himself very long to dwell on. He stuck the racket attachment out the window and now, with the mirror, he fished for the sun.

For as long as the sun stayed out, he practiced. The sunspot the mirror reflected was very small, about six inches in diameter, but he could see it on the forest wall clearly enough. He ran it down the road from the point at which the road entered the ravine, practicing to hold it steady at the one spot where the car would be briefly visible. He etched it up and down the side of the ravine, and back and forth. He practiced until he had near-perfect control of its reflection.

The Upper Carboniferous

Three hundred million years ago the land areas of the world were all of a piece, most of that piece lying within the southern hemisphere. Since then, although it has been possible at one time or another for plants and animals to move about over the world, it has not always been at their convenience to do so as the earth's hull eventually tore into fragments and floated apart, North America moving up and the largest remaining chunk, Europe and Asia, heading east. In the far south, where the ruptures were greatest, new fragments became continents, separating in an orderly sequential progression, one from the other: Africa pulling away from South America; Antarctica pulling away from Africa; and Australia pulling away from Antarctica. However tenuously, South America and Africa remained connected to the lands above them, the former tied by a land bridge at Panama to North America, and the latter to Europe by junctures at Gibraltar and Sicily. To the north, the Bering Strait held Asia to North America, and Greenland connected North America to Europe.

Even so, except for Australia, which drifted east and has remained isolated since, intercontinental travel has been manageable for plants and animals though only at irregular intervals. Continents move up and down as well as laterally; sea levels change. Sometimes the Bering Strait has been passable, sometimes not. The Panama bridge has been submerged for all but the last two million out of the past forty million years, causing an interval of isolation which accounts for many of the differences of South American animals.

Whatever the unknown circumstances weighing against the dinosaurs, they seem to have favored the mammals—which, although present almost as long as dinosaurs, had been understandably inconspicuous. Fifty-five million years ago, there was no mammal reaching even waist-high to a man (a species still fifty-two million years into the future). The horse then was the size of a fox; the camel the size of a rabbit; the elephant the size of a pig. Yet in going their own way, mammals had developed characteristics which lent themselves to a long-term run. With warm blood pumped by a four-chambered heart, they possessed a system of internal heat control, a thermostat against the weather outside themselves. Externally, they grew hair. They delivered their young live and produced milk to feed them. And they developed a brain larger in relation to body weight than any creature coming before them. Much of the expansion of this important organ occurred in the cerebrum, where memory and reason reside. They were more intelligent.

By trial and error mammals multiplied and changed, their errors heading off toward a weird bestiary of dead ends, a Jungian nightmare. There were camels with the neck of a giraffe, giraffes with the neck of a horse; there was a creature with the head of a horse, the body of a camel, and the claws of a cat. There were giant pigs; guinea pigs with long legs; sloths the size of elephants; elephants with tusks in their lower lip and some without trunks; bear dogs; a horse type built like a rhino but with a shovellike protrusion from its snout; a horse with horns near its ears and a unicorn's prong just above its nose.

Eventually, in response to the food supply, the larger land mammals began to sort themselves out, feet and teeth being two of their more significant distinctions. Plant eaters developed high-crowned teeth for grinding up vegetation and hooves for speed to escape meat eaters. Meat eaters developed claws to grasp and hold plant eaters and sharp teeth to tear flesh from bone. Some mammals moved into trees, developing toes and fingers to hang onto limbs. As prairies formed and the climate remained mild, mammals flourished.

Meanwhile, drifting slowly downward, the Antarctic moved into its present position in the south polar region, causing a change in the weather for the worse. Ice formed across the top third of the earth—from Oregon to New York, across most of England, over Germany as far as the foot of the Alps, and deep into Siberia. At any time the amount of water on the surface of the earth is a constant—most of it, of course, to be found in the seas. But glaciation in those times drew water from the seas and turned it to ice on mountains and over the land, lowering the sea level and thereby exposing old land bridges.

Whenever and wherever possible, animals moved to escape the cold, retreating from the advance of the ice, returning as it melted back. Deer and elephant came across the Bering Strait, passing horses bound the other way. The camel and the zebra wandered across Greenland. Llamas moved from North to South America; porcupines from South to North. Antelope, beaver, and the ancestors of domestic cattle crossed from North America into Europe and Asia.

The ice has advanced and retreated through four periods of glaciation, the most recent occurring 11,000 years ago. With each advance, the animals gradually were pressed back to more congenial settings, to places like the Sahara, which then, because of periodic rains, was savanna land filled with acacia trees and in some places jungle, the rich diversity of plant and animal life there having been recorded in Stone Age drawings. The great congregations of mammals then included elephant, gazelle, zebra, giraffe, hyena, lion, jackal, and so on, most of which may be seen today in zoos, or in special protected areas such as the remote African parks where they continue, within tolerated limits, to exist as they did during the Pleistocene epoch, two million years ago, when the extinction of mammals began.

Nairobi, Kenya

By now, in the late twentieth century, the Sahara has become the largest desert in the world. It extends from the top of Africa down to the Sahel steppe, about 1,200 miles; and from the Atlantic coast 3,000 miles across the continent to the Red Sea. It is one of the driest places on earth, and certainly it is the hottest—136 degrees Fahrenheit in the shade at Azizia, Libya, is a world record. There is little moisture on its surface, yet deep beneath it—under Libya, Chad, Sudan and Egypt—is a vast sea dating to the Pleistocene era of two million years ago, when the Sahara was a wet and verdant grassland plain. By the time

Rome was founded, the Sahara had become desert, desiccation spreading its edges and driving the refugee animals farther south; and the desert has continued to spread, burning across the remnants of African steppe at the present rate of three million acres a year.

At its edges—Lamprey, I know, certainly wouldn't put it this way —only a kind of tree impedes the Sahara desert's progress. This is the acacia—an angiosperm, like most modern trees, but nevertheless in every sense an unusual plant. Varieties of it are to be found elsewhere about the world, but it is especially abundant in Africa, occurring everywhere except in the deepest rain forests. There are fifty-five species of acacia in Africa. The most quickly recognizable is the loveliest of them, the tall, yellow-barked umbrella acacia. Its foliage is delicately filigreed, like that of the mimosa; bangled with the hanging nests of weaver birds; and flat-clipped abruptly and decisively across the top. Acacias are the sight and, in some species, the sound of the African plain. Ant colonies burrow into the round gall of their thorns, and the wind blowing over these small entranceways flutes a low melancholy call across the grasslands. In the heat of the African noon, the cool shade of the acacia is refuge.

The acacia plant in Africa is the animals' tree, and animals eat it in various ways, from bottom up and top down. The dik-dik, a deer-like creature the size of a rabbit, nibbles around the trunk. The gerenuk, a long-necked gazelle, starts where the dik-dik cannot reach and eats upward, stretching onto its hind legs to reach to eight feet. Bending low, the giraffe covers the largest belt of it, from four to eighteen feet, its higher range shared with the elephant which, growing impatient, may pull the tree down to get to all of it. Thorns and leaves are perennial fare, but the acacia's occasional and greatest delicacy is its fruit, a long, spiraling bean pod which, when ripe, hardly reaches the ground before it is gulped down by the traffic waiting below. Nomads feed their domestic stock on acacia fruit, and use the tree as fuel and for shelter; its bark has commercial value.

Men and cattle, antelope and elephant live off the acacia but its greatest value lies in its role as last defender of life, for plants and animals, against the marauding desert. At the edges of the Sahara, where the soil is still deep enough, acacias form the fall-back line, the last stand of savanna vegetation. With its spreading root system shaded by its own foliage parasol, the acacia stays green long after the grass surrounding it has burned out. Even before the rain comes again, it draws from the extreme heat to flower and turn its leaves green.

Where the acacia grows there is still some moisture left. The acacia is the tree of life.

Hugh Lamprey wouldn't put it quite so categorically, though so far as he is concerned it is certainly the most important tree in Africa—"*possibly* the most important tree in Africa," as he does in fact put it, the italics being mine. So is the wretched cold that began somewhere between Bombay and Karachi night before last, working from a scratchy throat toward the sinus cavity, arriving there at about the same time I hit Nairobi yesterday, in the splendor of a typically brilliant morning. Clogging up and threatening to descend into my chest, it is an annoying distraction to the appreciation of comparative distinctions. My bag is gone—on to Rome probably, since the connecting flight from Singapore to Bombay was Alitalia—and with it my prescription nose drops carefully included against just such a calamity. Invariably the cold comes in late August, the virus following the well-marked route of the year before, felling me on schedule by the third day, which means I had better get on with it now. Lamprey's schedule is no more flexible than my own. He has just returned from a walking safari, a 200-mile trek supported by thirty mules rented from the Masai, from Kekerok to Naivasha, and he has paperwork piled up.

". . . The species which we know as the umbrella acacia," he is saying, to be as precise as possible, "*Acacia tortilis.* The project arose about three years ago just as a thought. Very little is known about it. Man and his domestic animals are highly dependent on it, wildlife even more so. We are so used to thinking in terms of animals depending on plants but here was an example where a plant, and a very important plant in Africa, was to a very great extent dependent on animals." Immediately, as is his manner in conversation, Lamprey corrects himself. "It is a tri*ang*ular relationship, between the tree, ungulate animals, and insects. It seemed to me it would be rewarding simply on a scientific basis to make a study of this single species of tree." Now his study has become part of a larger rehabilitation project administered by UNESCO and centered on the area known as Marasabit in the North Frontier District; Dr. Hugh Lamprey is a director of the project.

We are on the fourth floor of Bruce House, in a conference room of UNESCO's offices, the window open to the soft air of Nairobi; down the hall someone is whistling the theme from *Dr. Zhivago.* An animal ecologist out of Oxford, Lamprey is an imposing man, in his mid-forties, in good shape from adapting to the shifting environs where his work takes him. In the last twenty-five years it has taken him where the flora and fauna of East Africa are most in jeopardy, most worth saving as a living corridor of time leading backward to the Pleistocene. He combs his hair flat, like a Welsh farmer, but he is got up in the snappy modern bush dress of the post-colonial Britisher: khaki shirt and slacks, bush vest, desert boots. It is the second time we have met—the first time was three years ago and for only a few hours—but I have some notion of what to expect. A slow talker, he has the unusual ability to form his thoughts into elaborate, often intricate sentences which fall into place as paragraphs—so long as he is not interrupted, which I am, under the circumstances, God knows, hardly disposed to do. To the extent he is able, Lamprey will tell what he has learned about the exchanges between plants and animals, to the critical advantage of both and to the even greater advantage of the African continent, all with reference to the exceptional case of the acacia tree.

Because of the importance of the tree to animal life—and because, as he came later to see, the acacia's being the last tree to resist the desert's spread meant that it ought to be sown, if possible, to push the desert back—Lamprey began studying the seeds of the tree to find out what it took to make them grow. What he learned was discouraging. The seed coat, having adapted to conditions of severe aridity, is hard and unyielding. No less stubborn than its resistance to destruction is its resistance to growth: germination occurs only after there has been an abundance of rainfall. In the African desert and well beyond it, an abundance of rainfall sufficient to draw forth the acacia seed occurs no more than once in every ten years.

Finally established, however, the acacia would be capable of looking after itself. Of this Lamprey had seen dramatic evidence. On a visit to southern Israel, he says, "I was taken round the Sinai by some Israeli botanists who showed me where *Acacia tortilis* had made a recovery. After repeated cutting and lopping by Bedouins, a few trees had survived as stumps in a watercourse delta in the desert, which held water once every ten years or so. They were protected inadvertently due to their presence in a military zone. They started to produce branches, and in about six years they were bearing fruit abundantly, carpets of pods three to four inches deep under each tree."

So the staying power of the acacia was evident; the sowing of seed in soil where it might grow was the problem. Undaunted by the prospect of having to wait ten years for sufficient rain and hoping somehow to induce germination under circumstances he could directly control, Lamprey collected seeds from under the trees before the hungry animals could get to them. And he collected them as well from the feces of animals that had eaten them. "The seeds are very hard, extremely resistant to crushing," he went on. "I noticed that the fruits, having been eaten by animals, are digested, but the seeds contained in them are not disturbed.

"Then we found a most extraordinary thing. In the Serengeti only a small proportion of the seeds which we collected from the pods in order to plant would grow. But we found that a very large proportion of the seeds collected from the droppings of these mammals *would* grow. The mammal acts as the dispersal agent by defecating seeds over the countryside. And not only does the mammal disperse the seed, it facilitates the seed's germination. At first we thought this was a well-known phenomenon of the chemical and abrasive effect of the passage of the seed through the mammal's gut acting on the seed to soften the seed coat and prepare it for germination."

But this was not what had happened. Upon closer examination of the unproductive seeds taken from under the trees, Lamprey discovered the third, and completely unexpected, side of the triangle: the seeds had been destroyed. "By insects!" he says, still somewhat bemused. "The insect concerned is a tiny beetle, *Bruchidius,* belonging to a group which specializes in laying its eggs on seeds, particularly of the Leguminosae. This particular species is completely specific to *Acacia tortilis.* It is dependent on this tree.

"The female bruchid hangs around the tree—we don't know what it feeds on yet—waiting until the flowers have dropped, and the fruit —the expanded ovary of the flower—has grown to the length of about an inch. At that stage the beetle lays its egg either on the surface of the fruit or just inside, where the seed is going to grow. The larva hatches out, and at a fairly early stage in the development of the seed —the larva is very small indeed, smaller than a pinhead—it burrows into the seed, and the seed then heals itself. The larva begins to develop slowly, feeding on the content of the seed—the cotyledons which are the food store of the seed.

"Now if that fruit develops in the normal course of events on the tree and then falls to the ground, and then stays there, the beetle larva will continue to grow and go on eating the inside of the seed. But it doesn't start to grow rapidly until the time the fruit actually drops to the ground. Until that time it is still small and has done rather little damage to the seed. The seed is still entirely viable. But once the fruit drops, the larva begins to grow very fast and to eat the seed so that within two weeks it has done a lot of damage. It may have eaten not only part of the cotyledons but part of the embryo as well, which is essentially the plant itself. So the seed is destroyed. About a month or six weeks after the fruit has dropped to the ground, the adult beetle drills a hole through the seed coat and flies away."

So efficient is the beetle that it is able to destroy most of the acacia seeds in the vicinity. "Now something else happens," Lamprey says. "As I've told you, the fruit of this tree is extremely nutritious to ungulate animals. And when the fruit is ripe, Thomson's gazelle, impala, elephant actually wait under the trees for it to drop, and they eat it almost immediately. So, under normal circumstances, the fruit does not stay on the ground for more than a day. You can go round underneath a fruiting tree for a week or two and find very little fruit there, because something has gotten there before you and eaten it. Now under those circumstances what happens is that the seed passes through the gut of the animal and in so doing the larva of the beetle is killed." He pauses and clears his throat. "So the passage of the seed through the gut of the animal is, if you like to put it this way, the means by which the tree protects itself against the beetle."

The digestive tract of domestic animals has proved as lethal for *Bruchidius* as that of giraffe and elephant. The nomadic Maasai people live among their goats and cattle, and in departing their settlement leave behind them thick manure spreads sprouting tiny *Acacia tortilis.* Since the process of preparing acacia seed is easier to manage with domestic stock, Lamprey's program to the north now uses goats within local communities for this purpose, employing youngsters from aid-assisted tribes to collect acacia seed from the dung.

But the acacia did not evolve its survival capacity in the long hot intervals between ten-year rainfalls through the benefaction of domestic goats. Moreover, it is well known among ecologists that the most ingenious human strategies are often undone by the "kickback" factor —the unforeseen distortions resulting from man's effort to rearrange a natural process. The Chinese destroyed birds to protect their rice and there followed an infestation of insects. To solve problems, the mutation of populations through natural selection may take epochs of recalibration, involving many more factors than goats and rain cycles.

If everything is connected to everything else, as conventional ecological wisdom now goes, then a single "solution" may not suffice. What can he realistically expect from his efforts? I ask Lamprey. "Can you outsmart the natural processes of the acacia tree, considering the circumstances under which it has evolved? Do you really think you can push the desert back this way?"

"We're dealing with a wholly indigenous species, and we intend to make use of it in its role as an indigenous species," he says. "It's still where it was but in ever decreasing quantities. One of the problems you have here is that through a very long period of degradation you lose the topsoil over enormous area. You have to face the fact that recovery to the original condition is a practical impossibility. So our goal is maximum recovery possible under the circumstances, which takes into account that only some acacia species are capable of recolonizing the country which they originally occupied, due to the lack of topsoil. In the sort of time span we have available you are going to get only a partial recovery."

"How long would it take for total recovery?"

"Tens of thousands of years," he says impassively.

An entomologist is coming to see Lamprey this afternoon and they will discuss the habits and life cycle of the *Bruchidius* beetle. Lamprey doesn't know yet exactly what it is that kills the larva in its passage through the ungulate intestine. The two of them will plan their strategies toward a future far beyond their own lifetimes.

The windows are open and the soft air of Nairobi is turning warm. My sinuses ache. For a passing moment I wonder what happened with Evans and his wildlife meeting.

In 1974, as a consequence of the Sahelian drought in which 150,000 people and many more of their cattle died, international attention was drawn to the accelerating spread of deserts around the world: within Peru, Chile, Argentina, India, Tunisia, and Libya (where the Sahara moves north), in all more than eighty nations. The causes nearly always are the same: the vegetation is cut down for fuel or timber and the topsoil is washed or blown away; or the vegetation is eaten by cattle and goats and then the topsoil is similarly dislodged. Just before the Sahara drought, protective measures for increasing the number of cattle were successful in the short term and fatal in the long term. Bore holes drilled into the desert floor brought water for the cattle but did nothing for the fragile vegetation, which soon was trampled to sand. (Water is a special problem: only 1 percent of the world's reserve is fit for drinking. In twenty-five years the need is expected to be four times as great as now.) Multiplying beyond the carrying capacity of the semiarid land, the cattle first began to die and after that the people dependent upon them. The vegetation there gave way to desert.

"There is a rather well-documented history," Lamprey says now, "of how the Sahara and Arabian deserts have expanded and taken thousands of years to do so, from minute beginnings of sand dune areas right in the middle of cities."

"Has agriculture always been involved?"

"Yes, agriculture followed by the next step, which is the overgrazing of livestock. It all began in a little patch of country where man started as a civilized being, an area near the headwaters of Mesopotamia about ten thousand years ago." He moves to a map on the wall and points to a spot between Iran and Syria.

I have with me a paperback textbook by the ecologist Raymond Dasmann, *The Conservation Alternative,* carried along because it is a clear and concise summary of issues relating to the interests of people I would be seeing. So villainous had the human animal emerged in Dasmann's view, however, that I found it necessary to discount many of his judgments until confronted now by the deadlines of Lamprey, a most restrained and temperate critic. I recalled a passage that seemed remarkably close to Lamprey's implications. Fishing it from my briefcase, I read aloud from it:

> Destructive changes in land and vegetation accompanied the domestication of plants and animals some eleven thousand years ago, at the start of the Neolithic. Until recently the change from hunting and gathering to agriculture and pastoralism was regarded as a major step forward for mankind —permitting peace and prosperity instead of the uncertain and brutish existence of the earlier age. This view has been challenged with convincing evidence, and some now regard the development of agriculture as the factual basis for the biblical myth of the expulsion of Adam and Eve from the Garden of Eden.

"Yes," Lamprey says after a moment. "I'm sure that's it."

7

A NEW PROBLEM

Trout Lake, Washington

"Please, God," he prayed, the same prayer he had prayed each day, out loud, "help me get out of this car so I can live." Then he coiled the wire into loops and began his chant:

Seven wraps of the string,
Throw as hard as you can . . .

He washed himself with the wet T-shirt and got ready for the day, Thursday, September 16, his fifth day in the hole. He couldn't get his eyes open all the way. At first he thought nothing about it. He had to get into his schedule. There was no point in getting impatient with the searchers, either. The Cascades were a whole lot of territory. His face felt funny, swollen. He looked in the mirror on the face of his tennis racket, and what he saw terrified him. His face had ballooned up; he looked grotesque. His eyes were narrowed to horizontal slits, like Eskimo glasses, so that he could only see things directly ahead. To see anything to the side he had to turn his whole head. He thought first that he must be suffering renal failure. He had worked with dialysis machines long enough to know what that meant. Your joints become inflamed; fluid enters your whole system, puffing up your eyes and your face. His joints didn't hurt, but his face was immense.

What could it be? Pneumonia? That was no reason for his face to swell. Maybe he had some sort of chest congestion, though it didn't hurt him to breathe. Or maybe it was the beginning of heart failure. He made himself take long, deep breaths to slow down his system. Could it have anything to do with his dead foot? He told himself to stay calm and not to start worrying beyond the inconvenience this new problem was now causing him.

But he soon realized it was a greater inconvenience than that and he could not just ignore it. He was able to fish in his water, to drink and bathe himself, make himself neat in line with his routine, but when he took the tire iron and twisted toward the rear he discovered he could see neither his foot nor the root which held it. He let himself fall down against the bump in the roof under him, giving way to the impulse to speculate further about his situation. It didn't matter so much that he couldn't see behind him to work on his foot. He had lost his foot. He had had so little success with the wrench, it didn't matter that he had to stop working on it. He wouldn't get out that way, anyway. He wondered about Mary and his mother, his brothers, and his two sisters. How were they? Was his mother okay? Where were his brothers? Where *were* they? He wondered what the swelling of his face meant to his chances of lasting until his rescuers came. If he knew what caused it, maybe he could do something about it. But what? He was doing everything he could think of now. *What* could he do? Send out for aspirin?

Well, *one* thing he could do was get off his butt and start bringing in water. Another thing was to get back to signaling for help. And another thing was to think about things he *could* do something about. Hauling in water, he thought maybe he should spend more of the time now available to him making himself comfortable. He tore up the upholstery about him and wrapped the cloth stuffing from it over and around his immobilized leg. Directly to his front he noticed that the spare tire was held by a butterfly nut. The jack was behind the spare. If he could just get the tire off somehow, he could maybe get at the jack. If he had the jack, he could jack the damn wagon off his foot —once his face had gone down so that he could see behind him again. He took one of the metal hot-dog sticks and began to flip at the butterfly nut. It went so far but no farther. He kept at it, between the times he tossed his T-shirt out for water, for several hours, until he decided it wasn't going to budge.

His air mattress lay a few feet beyond the wagon. Maybe he could make a pillow out of it. With his rod pole he managed to pull it to him, but he quickly realized it was pointless to use it for that purpose —there was no way he could ease his body within the confines of the wreckage. He tore out a large corner from it to make a receptacle for his urine. By relieving himself into it, he could get his waste out and away from the car frame.

He practiced maneuvering his tennis racket in various angles toward the road above. If the sun came out today, it would be available to him in midafternoon, and it would last only about three hours before it passed beyond the deep, narrow ravine. Until it came he had done everything he could do now except to keep practicing with the racket, and to fish in water. And wait.

The Eocene

The ozone layer, which protects all of life, is sixteen miles high. A mile down, life is found, and life is present thereafter, in one form or another, downward to the bottom of the sea. Life's movements are conditioned by lifeless substances—by weather, soil, fire, gravity, minerals, water, and the movements of land.

Forty million years ago, the hull of the earth began cracking along a vertical line from Russia to Rhodesia. It is still doing that, at the rate of an inch and a half a year. About halfway down the split, just below the equator, a chain of volcanoes rose a million years ago, throwing lava across the high, flat plateau to the west, the far side of what is now called the Gregory Rift. Over the lava there formed in time a thin layer of soil, too shallow for forest to grow but congenial to grass. Reaching from the Ngorongoro caldera in Tanzania to the eastern shore of Lake Victoria, the grasses covering this plateau comprise the Serengeti Plain, over which move still the last Pleistocene mammals, still following Pleistocene ways.

As the bison once dominated the American grasslands and the kangaroo the grasslands of Australia, antelope dominate the grasslands of the Serengeti. There are seventy species of antelope in all of Africa (whereas there are only two of rhino left, two of hippo, one each of elephant and giraffe, and three of zebra). The antelope have evolved in Serengeti because of the grass. Rain falls in abundance in the fall and the spring, and fresh new grass follows the rain over the plain, growing almost instantly after it falls. The promise of new grass puts the animals in motion, hundreds of thousands of them—200,000 zebra, 700,000 Thomson's gazelle, a million and a half wildebeest—moving in a great clockwise sweep about the plain. Wildebeest are said to be capable of sensing the rain's presence from thirty miles away, and of moving out ahead to get there when it falls. As the plant-eaters move, the meat-eaters—lion, cheetah, leopard—follow after them. They all move by evolved conditioning but in a collective fashion which suggests an ingenious system. Coming first, the zebra take the top of the grass. The wildebeest eat the leaves and the stem, farther down. The Thomson's gazelle nibble shoots of the herb layer. Meat-eaters take what they can get where they can find it, most often the lame, the weak, the newly born.

The antelope are in Serengeti because of the grass and most of the others are there because of the antelope. All exist in fluctuating states of equilibrium, living components balanced to a delicate tension. They turn in rhythm to the movement of the plain like the shifting axes of a gyroscope.

About the vertical axis: the light of the sun is converted by photosynthesis to produce glucose in the grass, which the gazelle eats to feed the jackal. In death, their wastes return to the soil to feed the grass, which the wildebeest eats to feed the lion.

Along the horizontal axis, where there is a different symmetry, the movement may sometimes slow to stop-frame sequence. Grazing near a marsh, a family of zebra prepares to move toward fresh grass. The path across the marsh is a narrow strand which will oblige them to go single file, a dangerous inconvenience. They bunch together, nervous and uncertain, listening, scanning the high grasses about them. One moves safely across the strand, and then another until all follow, some six of them strung out in line. Belly flat in the deep grass to their left flank, a lioness makes her move, charges, cutting one out of the file and within ten yards clubbing it to the ground with her paw. In seconds she is feeding. The zebra family regroups beyond the marsh, looks back briefly, moves on. Beyond the marsh a dozen hyenas assemble, and they begin to run in a wide circle about the lioness and the dead zebra, whooping and snarling until, in collective accord, they rush the center. Crimson-muzzled, the lioness gives it up, moving back into the marsh where she sinks onto the grass, unsated, awaiting another passage over the strand. The hyenas fall on the carcass. When they have finished the jackals will come, then vultures—all in a matter of hours. Then over days the arthropods will feed, and then bacteria and mold, the wheel slowly turning back to the vertical axis, as the flow of nutrients moves into the soil.

At the center of it all is the grass which, limited by surrounding desert, sets limits for the size of the antelope herds (which set limits for the numbers of predators). When the rain falls in abundance and there is more grass, there are more antelope. When the rain diminishes and the grass is sparse, there are fewer of them. Yet however dry the season may prove to be and how heavy the demand put upon the grass by the antelope still there, the grass will return in response to the rain of the season to come. Through an intricate, synchronous collaboration between the grass and the antelope, the one having evolved through natural selection to the need of the other, the wheels turn, the excess of antelopes dying of starvation before the meager grass remaining is destroyed. For the time being—Serengeti is protected by Tanzania as a national park—the Serengeti and its mammals go on, much as they have for millions of years.

Nairobi, Kenya

There are very few places in the modern world where mammals are left to themselves, still living according to the laws of natural selection rather than the interests of people. Mammals unfit for domestication, or otherwise valueless in material terms, exist at people's convenience. Between Evans's blunt pronouncement—"Man is an animal, there's nothing he can do about that"—and Ponnamperuma's more optimistic one—"Evolution is under our control now, we can direct it"—much seems to be missing in the mammalian precedent for man's own place within the laws of natural selection, laws which, according to Evans, he ignores at his own risk. Hugh Lamprey is an expert on the mammals of the Serengeti. A man of large projects, he was, before undertaking to roll back the desert, director of the Serengeti Research Institute, an international research station devoted to the study, maintenance, and preservation of the world's last great stock of Pleistocene mammals.

We were to have had dinner tonight but I beg off. I am certain I will not be ambulatory beyond sundown. Maybe he would just extend our session here in the UNESCO office instead, for there is much he can tell me, if he will, about the remarkable system of checks and balances that exists in the Pleistocene life about the Serengeti. Sequestered as a national park within an unindustrialized nation, it is also a model by contrast to the ways of modern animal life, unchecked and veering toward gross imbalance, just outside. Overgrazing of domestic cattle has brought desert to its borders; cattle die and men starve on the far side of the grassland plain. In Tanzania, the human birth rate is one of the highest in the world, and the nation one of the poorest. It is a disquieting fact of the nature of things that in Tanzania the last wild animals of the Pleistocene, following their own ways, seem to make do better than people. Through an understanding of the ecosystem there, some reasons for this cruel disparity become evident, and Lamprey, who has lived within and studied first hand the ecosystem of the Serengeti, can provide them.

The language of ecology, to begin with, is a bag of mixed metaphors and imprecise descriptions: the "food chain" is part of the "food web"; "diversity" means something more than variety, "carrying capacity" more than maximum load. The "pyramid of numbers" is mixed up with a concept called "biological magnification"; and it is difficult to know where an ecosystem stops and the biosphere begins. In the fullness of their ecological meaning, however, these terms comprise what Dasmann calls the rules of the game, and they force a view of the immensely complex state of living interdependence, the subtle and shifting relationships of plants and animals seeking balance. Would Lamprey untangle these confusions?

The biosphere, he says, is everything living on earth. There are relative discontinuities within the biosphere—where the seas separate lands, as an instance—and these discontinuities begin to define ecosystems, whole parts of the larger state in which there is a discernible exchange of materials and energy (although these parts, too, are invariably connected in one fashion or another, by wind, or rain, or rivers, or by animals crossing them).

The sun runs the ecosystem, working upward through green plants or algae into the food chain. Ecosystems can be of any size and any degree of inclusiveness. A drop of water has an ecosystem; so does a cow pasture; so do the Black Sea, New York City, and the state of Nevada. Ecosystems are biological communities. They consist of plants and animals living in community, as well as the climate, matter, and energy necessary to hold them together. Since life is interconnected—any skein of it leading off to innumerable other skeins—it is therefore not always clear where one ecosystem ends and another begins.

An ecosystem (Greek root: *oikos,* meaning "house") is not fixed in its proportions nor along a set time scale. It changes. The most self-contained of ecosystems—a "climax" ecosystem—is one in which a community of organisms will endure indefinitely unless disrupted by some natural (or unnatural) phenomenon. Though a drop of water is a working ecosystem, it is of limited duration. The Serengeti Plain in Tanzania (about the size of the state of Connecticut), or part of it anyway—the grazing grassland—is still a climax ecosystem, as it has been since before the beginning of the Pleistocene.

So far as I can see, the ecosystem is at the heart of all of it. I ask Lamprey: "Is it possible to know all the parts that make an ecosystem go?"

"It's a bit ambitious to imply that we could know everything. You've probably heard of the incredibly simple ecosystem created in the laboratory in a tank of water into which you put some grass and just let it brew for quite a long time. The breakdown products of that grass, along with a few microorganisms which go with it, will create a self-sustaining ecosystem dependent on the nutrients present and the atmosphere in contact with the water surface. But nobody has kept one going indefinitely because it is deficient in certain elements which man is incapable of putting in in the right proportions.

"With ecosystems you've got to realize that you're dealing with systems very far from fixed in their proportions. The interactions within them are dependent entirely on feedback mechanisms which keep the whole thing within certain limits. These processes will take place at a certain rate—expanding energy, turning over energy and materials—and all of them will fluctuate about a mean. If the processes start to depart from that mean by a relatively large amount, then among the elements feeding into the system there will be produced an effect which becomes progressively more limiting."

Lamprey pauses a moment and thinks. How to put it to a nonecologist? He comes up with an example he assumes may be closer to my world than his. "It's a bit like the internal combustion engine and the simple feedback system of the carburetor. If the carburetor is correctly set, the engine will idle at a set speed. If it's slightly incorrectly set, you get a process known as 'hunting,' a fluctuation in the speed of the engine. Petrol is getting in faster than it should, causing the engine to speed up. Eventually it will reach a point where another jet in the carburetor will take over, limit the speed, and bring it down again. So an engine which is incorrectly set will go up and down. You notice this very quickly in diesels.

"What a well-regulated ecosystem will exhibit is a very even turnover of energy and materials. Numbers will tend to remain stable, but you can only say that this will happen in a stable environment. It can't happen in an unstable environment—due to weather, primarily, and to some other natural causes. But now we have one other major influence. Man. Before man became a major influence on ecosystems, more than just an element within them, ecosystems had the capacity for limited self-regulation provided they were not displaced in any direction too far. Having got to this stage"—he is suddenly beginning to look a bit haggard—"I strongly recommend that you start reading the introductions to some of the academic ecology texts. I suggest this because I believe you're getting yourself in so deep that you'd better go into some of the books by people like Howard and Eugene Odum." Howard Odum, I later learn, is prominent among those extending the principles of ecology into other fields, especially where the intersection is inevitable—the field of economics (whose Greek root, Odum points out, means the "management" of the house).

"Food chain" ought to be obvious enough, Lamprey says. It is simply a case of organisms feeding upon one another, all enmeshed in varying connective links with the energy of the sunlight providing the first level of the food supply. But the conversion of energy to food and its various labyrinthine ways across food webs are more complicated. Certain nutrients in the food, which come from rock or water or air or soil, may move around a single ecosystem for thousands of years, an atom present in the first living creature moving through time into dinosaurs and still circulating today. For example, an ion of sulfur extracted from soil and resting in a blade of grass may be consumed by a gazelle to become a protein component of gazelle muscle, thence (when the gazelle is eaten by a lion) to lion muscle. Once the lion dies, the near-immortal ion is consumed by reducer organisms which eventually return it to the soil, to surface again through a blade of grass. Minerals depart an ecosystem occasionally through natural causes—with the wind, by flowing water or traveling animals—but they are rapidly lost in large quantity through severe soil erosion, and with their departure the ecosystem loses its capacity to sustain life.

But it is solar power that drives an ecosystem, the energy carried through green plants into food chains, which may be seen as series of organisms literally eating each other up. At each step in the food chain 90 percent of the energy in the previous step is lost, the attrition moving upward through the "pyramid of numbers," which Howard Evans, from the vantage of his back porch where he reflects on the innate superiorities of the bugs that bemuse him, has described as

> a very great number of very tiny organisms, a slightly smaller number of slightly larger ones, a still smaller number of still larger ones, and so forth: each level of the step pyramid depending very largely on the one below it. On the top stands man and a very few other truly large organisms, and in a sense man's brain places him in a lonely and well-fed eminence above even the largest. The tragedy is that we on the top are so large that, even with our remarkable brains, we have difficulties understanding the creatures on the bottom—and very little time for this sort of thing—though without them the whole edifice would come tumbling. The pyramid of numbers is also a pyramid of knowledge (or ignorance), for what we know about an organism is often roughly proportional to its size.

Of all ecologists' terms, surely the most inadequate is the word "diversity," which by conventional definition only means variety. Yet "diversity" carries much of the weight for justifying ecological planning, the full comprehension of the term being essential to any argument that a dam ought not be built if it disturbs the habitat of a unique salamander downstream; or that wilderness area should not be mined for coal; or that the Serengeti ought to be preserved at the expense of the needs of hungry people seeking more grass for their starving cattle. It is too weak a word for all these tasks.

In ecology, diversity means the complementary variety of species of plants and animals within ecosystems, and to ecologists this in itself has an inherent and vital value. I asked Lamprey why.

"Diversity tends to promote stability. The more species that you have interacting with each other, the more stable the ecosystem will be—the more difficult to disrupt it, the less it will fluctuate." Species diversity and biological productivity are evidently interdependent. Chinese fishermen, coming now to be seen as the most advanced fish husbandrymen in the world, operate on the principle that the more fish species there are in a body of water, the greater the gross of production. There is more biological productivity in the tropical rain forests—more species of plants and animals interacting—than in alpine climates of the northern hemisphere; hence, more of the stuff of life. A single volcanic mountain in the Philippines has more species of woody plants than are to be found throughout the entire United States; to this degree, its ecosystem is more secure, and this is something to think about.

"Maximum diversity is what you aim for when you set up a national park," Lamprey says, as another example, which is closer to his own experience. "It is one of the more desirable characteristics of an ecosystem because it assures stability. We recognize that the earth is populated by animal and plant communities in different places, each of which is the culmination of a long period of evolution to its present place with its own characteristic diversity. One of the first things that happens in an ecosystem that is experiencing even the beginnings of disturbance and resulting degradation is loss of diversity."

"This can happen very quickly?"

"Yes."

"Because of the interdependence factor, as with your acacia tree?"

Lamprey nods and comes at it from another way. "The *antithesis* of diversity is what man does with his agricultural monocrops. He replaces diversity with its opposite. He will take an area that formerly supported two or three hundred species and replace them with one. This produces an instability which can only be counteracted by a highly developed agricultural regime—enormous effort, energy, expertise to go into it—to maintain the crop. The monocrop is extremely unstable and it hasn't the slightest chance of self-perpetuation. *There* is the antithesis. The less diverse the ecosystem, the more unstable. And the lack of diversity has several different implications. One is that you present to some lucky predator, some pest, the feast of its life, the opportunity for its population to explode.

"But this is not so in a diverse natural habitat," Lamprey goes on, "where every species has to seek out its own food plant, its own host plant if it's a parasite, its own special source of food. Nothing is easy. There is a struggle. To suddenly present a predator with a pure strand of sheep unprotected as you have in parts of New Zealand or Australia —I've gone too far on that, because your next question will be, what about the wildebeest in Serengeti, and isn't that a monocrop? It isn't, in fact. It is just an area in which one species has happened to predominate. This doesn't take away from the fact that the diversity is still there among other species in slightly fewer numbers." I am ready to move on but he isn't. "Another thing about diversity is that it is characteristic of a species to be present in certain numbers in its natural habitat. And those numbers—density, if you like—depend on the resources upon which that species depends. In the Serengeti, the fact that the wildebeest is numerous and the eland far less numerous is a simply explained fact. It's because the wildebeest's food is the grass, which outnumbers every other plant type there, ten to one or so. The eland's food is the broad-leaved species which are found growing among the grasses. It comprises *less* than a tenth, perhaps a hundredth, of the total food source available.

"If the food source is there, there will be an animal that has evolved to eat it." It is a striking statement, and it hangs in the air before he moves on. "And the characteristic size of the population and everything about that animal which is adapted in relation to size of population are due to the abundance or rarity of the food and other resources it depends on."

Does that relationship, I ask Lamprey, have to do with the term carrying capacity?

Thus far, there has been an equanimity about the man difficult to disturb; this question seems to have breached it. His eyebrows slowly ascend. "It surprises me very much that the concept of carrying capacity is so little understood, because it must seem to most people, upon the slightest reflection, that it is the most logical point you could possibly think of. Resources are finite, in any place! And the capacity of those resources to support consuming organisms *must* be seen as limited. It is the most surprising thing in the world that anyone could not only miss this point but not think of it for themselves. There is a limited amount of food. Food isn't conjured up out of nothing. You can only grow so much wheat on an area of land. Therefore you can only feed so many people from that area. That is carrying capacity. What more need be said?"

On behalf of others like myself, which I hazard to be the majority, I suggest that in the West—in America at least—consumption has been based on such an abundance of resources that people do not instantly think of cycles of planting and growing, or that farmland ever could run out. Moreover, given the technological advances of recent years enabling the quantities of agriculture to increase as against the fewer people needed to manage them, and with more and more people moving into the cities . . .

"I'm sorry," Lamprey says, genuinely abashed. "I made what is probably a basic mistake. The prairies of Canada and the U.S. are a special case—an enormous special case. But in those areas which are habitable by man, very many of those areas are already inhabited to capacity. The great edaphic grasslands of North America and possibly a few others—in Australia and Argentina—are an enormous special case. And that capacity to produce surplus food is what is keeping a large proportion of people in the world alive today. But carrying capacity in its more general form is just as obvious as I said it was."

Could he give examples of carrying capacity in the Serengeti?

"The eastern Serengeti grasslands are apparently adapted to what we would call a grazing ecosystem. They are as they are because over hundreds of thousands of years they have been grazed by wild animals. There is one grass species that occurs only in that area and is possibly in response to heavy grazing—grazing itself, within certain limits, tends to stimulate growth. This grass exists simply as a mat on the surface rather than as tall stems."

"It's protective?"

"Protective? You've got to be careful not to be teleological. It holds the ground together." From a recent count by the Serengeti Research Institute, a million and a half wildebeest were found to be existing on these grasses, the dominant animals in the area. "The equilibrium level is reached," Lamprey continues, "when the wildebeest and other grazing animals are in balance with their food supply—the grass available." To a visitor in the area, the importance of this balance is quickly evident. On the same day one may drive past the flourishing herds of wild antelope across the park borders into rangelands where dead cattle are festering, bloated, with their feet in the air. It is this odd, dramatically shocking disparity that begs more explanation.

"The beginning effect on the grass, which affects the grazing species, then, is the amount of rainfall?"

"Yes. We have a formula for it. Generally, the relationship is that one kilogram of grassland vegetation per hectare per year will be produced by one to two millimeters of rainfall. And that is a most simple and useful concept. It gives you a numerical value for what will grow and the limitations on it."

"Serengeti is a restricted area in that it is a national park?"

"Yes."

"Why wouldn't you have the same factors at work if you had a similarly restricted area for cattle?"

"You would."

"But will cattle grow in number in similar circumstances to the wildebeest?"

"Yes, but with certain qualifications. The carrying capacity of grassland under cattle has another dimension to it. That is management, and management will radically alter the carrying capacity you can expect of that grassland. In the near-natural situation of the Serengeti, management is achieved not by man but through a very complex adaptive situation which is free to act, to go on reacting within all the elements which do react with one another in an unrestrained way. Serengeti has rather few human-imposed constraints upon it. It continues very much as it always was."

"What is the nature of the management that causes cattle to respond differently to the same grass?"

"The first thing is this: you can select just so many cows to put onto your fenced area . . ."

"The number of cattle hasn't *grown* to that limited area?"

"That's right. The process hasn't evolved by itself. The farmer knows, either intuitively or because he's measured it, how many cows he can put on there, how long he can put them on, and he knows the results of that degree of intensity of grazing. Then he also knows that by putting those animals on that piece of land he is taking energy and nutrients out of it. If he continues to limit the utilization of that piece of land to its own capacity to regenerate nutritionally, then he needn't do any more to it—just leave it, provided that he maintains the regime at a particular level. If he wants to increase the intensity of that piece of land, he's got to put something back, and then he will start adding fertilizers to it."

"So the carrying capacity is no greater for wildebeest than for cattle?"

"It probably is greater for wildebeest, because it would be very surprising to find a farmer whose skill would match the adaptive situation that has evolved."

"Because the wild species is constantly adjusting itself to these circumstances?"

"Yes."

"What do you call the controls that allow this to happen?" A clumsy question and I am very much aware of it but I don't know how else to ask it. In the hall outside the office, the whistler starts up again. It is still *Dr. Zhivago.* My cold now is definitely in my chest. I can hear it down there.

"Natural selection is the main control. That is the selection which fits the organism to survive and reproduce best in its environment. You've probably read of the struggle for survival and the survival of the fittest. Now you can look upon this as a struggle not only of the animal species to survive in the face of a limiting environment—the main limitation being food—but also the struggle of the plants to survive in the face of the pressure which is being put upon them by the animals.

"When I say plants and animals, this single interaction is only a small part of the web of interactions we're talking about. It's going on *all* the time in *all* the species within this web of interactions. Each species in turn is being constantly refined genetically to fit into that particular niche. And because of the multitude of interactions you are getting a *highly* refined state of equilibrium.

"But when you raise the example of a farmer trying to achieve equilibrium, you have to remember this: the Serengeti remains self-sufficient because it is an almost totally recycling system. The animals are born, they live, and they die there. The Serengeti owes its ability to go on unchanged within its own fluctuating limits to the fact that it is a recycling system. Now, human farming practices are not, in the same way, recycling.

"Every time a cow is raised it is eventually eaten. The milk product from that land is taken off. This is a very different matter. You are carrying out what some ecologists call subtractive processes. You're taking away from it the whole time, and this is exactly what is going on in much of the grassland of East Africa under animal husbandry of varying degrees of sophistication.

"Every time a cow is sold in the market, that represents a subtraction from the ecosystem. Another side to this is that the wildebeest are dispersed over their area—they move across it. They distribute both their consumption and their output. The Maasai have to bring their cattle inside a fence every night; so for just about half the total hours of the day, those cattle defecate in one place. This means that something like half of that part of the recycling of nutrients is denied to cattle-grazing land; and no effort whatsoever is made to redistribute it.

"In contrast, on the average farm in Europe, the farmer makes a point of muck-spreading. He collects manure from the farmyard dungheap, takes it out in a cart, and spreads it over his pastureland. In that way a part of the cycle is kept complete. And of course some farmers go a great deal further and put down other materials as well, which may or may not be suitable for restoring grassland. But at least they are putting something back.

"Now the greater part of the semiarid grasslands of East Africa are undergoing continual decline due to (a) overgrazing, which we don't have to define—just too many animals over and above the carrying capacity; and two subtractive practices, (b) taking the animals away and (c) removing half of the dung which should have gone back."

Yet the cattle in Africa continue to multiply despite the erosion of grasslands and the spread of the Sahara southward at the incredible rate of three million acres a year, a domestic population thriving in the face of diminishing resources. To account for this paradox, Lamprey turns back to the weather.

"The drier the country you're in, the more the climate fluctuates within that area. It is a characteristic of the arid zone that it is also highly irregular. So one of the things that happens is that after a very good year—the one good year in ten, with plenty of rain—you get a tremendous superficial response in the vegetation even though you've had degradation in the past. With good rain you get tremendous annual and ephemeral growth."

"By ephemeral you mean temporary?"

"Yes." He explains that after such a rain, which produces food however ephemeral (and the germination of acacia seeds), the people who own the cattle take advantage of the opportunity to let their herds expand, and nearly always they do so to the maximum. Within two years there is a doubling of the cattle population, and when the dry phase follows as it always does, the land once more is overpopulated. Cattle are sustained somewhat beyond natural deprivations of famine and disease by hygienic controls and by certain emergency measures, but overexploitation of the land increases exponentially. I ask Lamprey: "Doesn't death by starvation eventually become a limiting factor?"

"Yes. Well, you see, we talk about thresholds of limitation—there is a different threshold which acts at a slightly different population level. That threshold in the wild just happens to be a threshold which permits the overall survival of the system. Except under the most unusual circumstances, animals in the wild reach the threshold of population increase or mortality *decrease,* at a level at which they are not damaging their environment." I didn't realize it at the time but this was the most important thing he had to tell me. At the time, I was still trying to understand his terms.

"If they do go over that level, is there a population crash?"

"It varies by species and circumstances. You recall our discussion of feedback? If the population for some reason increases a long way above the carrying capacity of the land, the turndown will be a crash. If it only rises a short way, the fluctuation tends to be a minor adjustment."

"But in a crash does it fall dramatically below the level at which it had been? Well below the carrying capacity?"

"Yes. You have to look at some of the documented examples of crashes, and the best of these probably is that of the kaibab deer herd in the western United States. The data have been variously interpreted, and the whole thing has been argued rather thoroughly, but one interpretation is that a very important element in the ecosystem was removed, and that was the predator—the wolf and the cougar which fed on the deer. These were not regulating factors, they were *marginally limiting.* This is the really important, the *crucial* point. How are animals regulated? Regulation among ecologists has a rather exact meaning. Regulation carries with it the meaning of the feedback system, a direct relationship between the numbers of the animal and the environment, and a regulating mechanism which tends to bring that population to equilibrium with the environment.

"Whether you call it a crash or not depends on how sudden the decline is and how far it goes. Populations of animals in temperate zones that have been studied, primarily the snowshoe hare of North America and the voles of Europe, go through fluctuations in numbers. Due to breeding characteristics of the population, they will tend to expand in numbers up to a point, and they will then rapidly decline. With voles and snowshoe hares, this cycle tended to have a certain predictability about it. The cycle was found to depend to a very great extent on the food supply but with the added element of predation, which would simply truncate the top of the peak. Although this has been extensively argued because it is a biological principle of vast interest, we don't have a certainty of how it works in detail. We *do* know that it works. The assumption that has been made by the best predator biologists is that populations tend to fluctuate in response perhaps to triggering effects in the environment, while the predators, although not exerting a great pressure, simply tend to flatten out the fluctuations.

"So when the predators were removed from the kaibab deer, this flattening out of the populations was removed as well. In a particularly good year for food, instead of the peak being reduced, it continued to rise. And then it continued to rise beyond that. Perhaps predation was more important than anyone understood. And *this* was the important point.

"Eventually, while we can say that food supply is the ultimate controlling factor, under some circumstances the food supply will go on supporting a population to the point where that population is eating more of the plants present than those plants can sustain in the long term. To put it another way, you can eat just so much of the plant and it'll be perfectly all right—a perennial plant like a bush. If you eat more than a critical amount, that plant begins to die back. And if you go on eating, it will die altogether.

"We have a certain amount of information on what this can mean. For instance, it appears that the short grasses of the Serengeti Plain can sustain consumption by wildebeest and other animals in an amount which is somewhere in the region of 70 percent of the whole growth that takes place, provided that the rest is left behind for the plant to feed itself through photosynthesis."

"Otherwise you get desert?"

"When you overstep that limit, yes."

"Should the wildebeest population fall to half its size—to seven hundred fifty thousand—would their birth rate adjust in some fashion to the diminished food supply?"

"You mustn't confuse two effects. One is that the actual proportion of calves to females born tends to remain constant under conditions observed so far. The other is that there is a differential mortality in calves depending on the conditions during that year. There is a tremendous flexibility built into animal populations in their ability to increase in number provided by this great reproductive excess."

The light is beginning to dawn. What he is saying is perfectly consistent with Darwinian principles. I should have grasped it much sooner. "The limiting takes place *after* the reproduction process?"

"Yes, that's it. The birth takes place, then there is mortality of the young. This has been true of human beings as well. Most human beings in uncivilized societies die in their first year. This of course is changing now and is one of the great causes of our problems."

As with the Pleistocene wildebeest, so presumably with people.

8

HELP FROM OUTSIDE

Trout Lake, Washington

Well, there she was, thank God, good old "Bert"!—Dr. Alberta Gabroccio, who knew all about renal failure. It was okay now. He rested in the litter, all his weight distributed at last on a horizontal plane, and Bert was in charge, giving the orders and making damn sure Vihtelic would be all right. Two ambulance attendants were moving about and one of them came up to him and said he could just take it easy now, they would have him out by tomorrow morning. So Vihtelic went back to sleep, and when he woke up—it was Friday, the sixth day—he waited for them. He waited through ten, then till eleven, and then on through twelve. After that, he resumed his routine.

Later—the next day, Saturday, the seventh day—Val came over. He had been picnicking with Elena and Elena's two little girls. They had all their food and stuff at one of those concrete picnic tables. Must have been a dozen of them. The little girls were giving Val a very hard time but he was very good-natured about it. He really took them seriously, even though they were Elena's children and not his. And suddenly there Val was outside the station wagon, very surprised and concerned to see Vihtelic pinned inside. Val told him not to worry, he was going right away for help.

Then the next day, which was Sunday now, the eighth day, Bruce Duncan came by. How did Bruce know he was there? Maybe Mike Jenkins had told him; Mike and Vihtelic had said they were going to Rainier, and Mike was the one Vihtelic had worked out with Thursday and Friday, running after work to get in shape. Mike had stayed back, and it was probably Mike who had told Bruce, and now here Bruce was, bending over, telling him the worst of it was over.

"Okay," Bruce said, "we've found you now. We can't get you out of here. We'll have to go get an ambulance. We'll get something to eat first. And then after that we'll call you an ambulance."

Thank God, Vihtelic thought. "Just let me sleep until the ambulance gets here," he said to Bruce. He felt very good now, relaxed and self-assured. He was through fishing for water. Very soon he would have all the water he wanted to drink any time he wanted to drink it. He could have ice water, as cold and clear as the little fountain of it fifteen feet away that had teased him through the past week.

He asked Bruce if he would make some calls for him. Bruce said, sure. There would be three in all, Vihtelic said, one to his mother, one to Mary, and one to Mary's parents. All of them would be worried, and he'd like them to know he was all right. Bruce said he would be glad to make the calls.

Vihtelic had a pencil he'd found in the glove compartment. He took a calling card from his wallet and wrote each of the numbers on the back of the card and put the card on the windowsill. Vihtelic said okay, he would just go back to sleep now while he waited for them to get the ambulance. He slept a long time and quite well, and when he woke up, he looked closely at the card, which was on the windowsill where he had left it. The three numbers were there just as he had written them although two of them were wrong.

Vihtelic waited for Bruce and the ambulance until four o'clock, and then he began to cry.

The Paleocene

Seventy million years ago, before dinosaurs died, before the earth split open down the African continent and angiosperm grasses spread, attracting antelopes to graze, the first primate emerged—a tiny, inconspicuous night creature with claws and a long snout. It looked like a rat and ate insects. True rodents, in fact, were its chief competitors for food; losing out, the ratlike primate moved into the trees and fed instead on plants.

There is nothing left of it today, although an immediate descendant, the tarsier, survives in the East Indies (and only there). The tarsier is unsettling to observe. Its head is disproportionately large for the size of its body, its eyes disproportionately large for the size of its head. Unlike those of most other mammals, the eyes are set forward rather than along both sides of the skull. Its arms and legs taper into long fingers with coverings more nail than claw. It has a long tail. The tarsier is an atavistic echo of a turning point, beguiling to regard until recognition turns it to caricature, then slightly grotesque. It is a tiny animal; you can hold it in your hand. After the tarsier, monkeys and apes arose, all of them, like their progenitor, dwellers in trees.

Living in trees is hazardous. Leaves block the view; branches can break, and the force of gravity punishes a false step. Space and distance must be calibrated precisely; when an error is made, the adjustment must be instantaneous. The eyes of early primates moved forward in the skull to provide three-dimensional vision for more accurately judging distance. Within the brain, behind the eyes, a network of nerves grew sensitive to color. The claws evolved into hands for grasping. Living in trees changed primates. The tarsier reflects some of these changes, which brought still others, leading to consequences that would extend far beyond the first occasion for change.

Thirty-five million years ago monkeys arrived. They perfected the art of living in trees. The world about them was neither flat nor monochromatic but proportioned in space and distance, filled with shadowed forms, textures, and color. Unlike the first primates whose fingers moved together as a single unit, this new primate, the monkey, could move each finger separately. The monkey could hold things, turn them in its hand, smell, and examine them—could separate an object from its setting, could abstract it (an achievement that in later primates would lead to language). The cortex of the monkey brain, obliged to coordinate muscles responding to unpredictable circumstances, expanded in size and changed in structure. The task of surviving in trees fed back the beginnings of primate consciousness.

Although monkeys leaped and rested in the vertical position, they traveled across limbs on all fours; their food was leaves. Apes arriving after them (the earliest fossil ape is dated at twenty-eight million years) sought instead the fruit of trees, which grows at the ends of limbs where in the interest of future trees it is less accessible. To reach the fruit, apes swung through the limbs, developing greater power in their shoulders, arms, elbows, and wrists. Thus did diet change the locomotion of primates from the very first down to the most recent.

Seeking more food, apes increased their range. (Monkeys tend to remain close by; apes can travel for miles.) As possibilities increased for the ape, so did choice. The ape began to bide its time, to delay, to watch for the best opportunity, to select one course out of many. The ape began to learn from experience.

As the climate of the world cooled and the forests shrank, as the grasslands spread, inviting the antelope and their meat-eating dependents, some apes began to come out of the trees, themselves attracted to new food opportunities, and they brought with them the new skills they had acquired away from the ground. The ground is a setting modern apes are not yet thoroughly at home with. The gorilla and the chimpanzee move across it awkwardly, tending to scrabble in a half-squatting position, leaning forward on the knuckles of their long arms and pushing themselves ahead by the sides of their feet. They are capable of standing upright and walking that way, but not for long. Their skeletal structure does not permit it.

From among the apes feeding at the edges of the grassland plains, rising tentatively to peer about, a different creature emerged. It was about three and a half feet tall, near the size of a modern pygmy chimpanzee. Very little else is known about it since the evidence remaining consists of a handful of bone scraps. They date to fourteen million years ago, perhaps earlier. The name given the new animals is *Ramapithecus.* It derives from Rama, for the human form of the Indian god Vishnu; and from *pithecus,* the Greek word for ape—god ape.

Cambridge, England

The soft rain is still falling, and I have two hours on my hands before meeting Thorpe across town at his quarters in Jesus College. The driver lets me off at a pub, and I take a table between the electric heater and a bay window, order a coffee, and dawdle. His presence has never been very far from the concerns that set this improbable itinerary, and it is therefore scarcely worth noting the improbable nature of the place in which I wait to see him, with my improbable questions. Dawdling still, I note it nonetheless and decide it is appropriate enough, as appropriate certainly to the occasion as any of the others I have visited.

Three years ago I had never known of the man, but through the coincidences which have brought me here he has become decidedly more important to me than the others. The question I had carried home from the Serengeti—how could it be that wild animals were more successful in their use of resources than intelligent men?—required the reflections of what sort of specialist—or who in particular he or she might be—I couldn't have guessed. Scientist or humanist? And then in the library, along that part of the biology shelf given over to the rather recent study of ethology, or animal behavior, I had browsed my way to the quiet but telling title, *Animal Nature and Human Nature.* Right away in his introduction, as calmly as his title promised, Thorpe pointed a better way at least to *perceive* my question:

> It is only too clear, and not in the least surprising, that human life appears to be less generally valued than at any time in the past hundred years or more. There are a number of possible reasons for this, but two of them are I think particularly clear. Firstly there are already too many people in the world so that, however humane and benevolent we may be, one cannot help feeling a twinge of satisfaction (if the disaster is not too near us, personally) at anything which reduces the population a little, however trivially. Secondly the widespread, often subconscious, acceptance of the outdated view of science, that man is simply the product of vast machinelike evolutionary forces, can easily lead to a hopeless loss of faith in the value of human beings and a consequent sense of the futility and purposelessness of our existence. It accordingly seemed to me that the position of Man in the animal kingdom, his evolutionary relationships to the rest of the living world, as far as these can be reasonably presumed, is one of the most important and significant topics to which a biologist can address himself. And granting this, it is essential that people should be shown both the characteristics in which the animal world approaches, and in some cases greatly exceeds, mankind in achievement and those characteristics of man that cause us sometimes to feel that we are brutes and at other times that we are gods.

To the end of setting out his own perspective and without moving from his writing desk (so far as I could tell), Thorpe then went on to survey much of what is to be found on some of the other shelves that surrounded *Animal Nature and Human Nature.* What is the significant difference between living and nonliving matter? "Doing something," he decides, citing Schrödinger: any piece of matter that moves, exchanging material with its environment, is alive. How does life come about? Despite the recent achievements of geneticists, nobody knows, or is likely to, and he states the frustrating riddle succinctly: *"Life is not merely programmed activity but self-programmed activity."* Nonetheless life has advanced, and with no declared motive beyond his intention to compare and contrast ourselves with all others, he proceeds skillfully to track its progress, synthesizing the "chance" theories of Jacques Monod, Kepner's concept of a group mind among the cells of the body, quantum theory, dualism—from Descartes through Popper—and on and on, with uncondescending consideration for the uninformed reader, and with undiminishing wonder. On molecular genetics, for example:

> The layman can gain much from a vivid simile which likens the information store in the DNA chains to instruction books which can be closed and put away or opened and read out. Thus it has been suggested that a bacteriophage, or virus, with a DNA chain, say 200,000 bases long, has in its molecular instruction book 60,000 words, which would be roughly thirty pages of an instruction book written in English . . . .

From the storage of information in simple organisms to the advent of communication—the exchange of information in higher animals—Thorpe works forward to the critical distinctions between animals and men. Not tool-making as once was suggested (bower birds use twigs to build their bowers and then dip their beaks into plant dye to paint them); not language (apes respond to men through sign language); nor perception (animals of all sorts perceive more than we can at present account for). With Whitehead he concludes that "the distinction between man and animals is in one sense only a difference in degree. But the extent of the degree makes all the difference. The Rubicon has been crossed."

It is within the mind that this great difference in degree lies, Thorpe says, and while the precise whereabouts of the mind inside the brain are uncertain, its enormous resources exist as well outside itself—indeed, outside the body—in the world's libraries of accumulated knowledge (in science, history, literature, and so on). Thus, while it may be said that animals learn from direct experience, man learns as well from the experience of other men who have gone before him. The record left—culture—is the quintessential intangible asset, the heritage of humanness—the great difference in degree. (It is this spur, cultural evolution, which Evans so casually suggested has changed humankind, and will continue to change it, at a much swifter rate than natural selection.)

So if it is by the act of thinking that man has acquired his store of knowledge, Thorpe turns back now to the animal kingdom to see where *thinking* came from—to explore the origins of consciousness, tracing them through research into animal behavior forward to higher orders and ultimately to the phenomenon of human perception, the catalytic agent of thought. It is a brave and noble search but ultimately frustrating: the processes of thought are not yet powerful enough to understand their own nature. Yet what a thrilling climax to the advance of life they represent! Thorpe draws on the eloquent neurologist Sir Charles Sherrington for his account of what is known of the ineffable sequences when the body's energy delivers to the mind an object for its contemplation:

> For instance a star which we perceive. The energy scheme deals with it, describes the passing of radiation events into the eye, the little light image of it formed at the bottom of the eye, the ensuing photochemical action in the retina, the trains of action potentials traveling along the nerve to the brain, the further electrical disturbance in the brain. . . . As to our *seeing* the star, it says nothing. But to our perception it is bright, has direction, has distance. That the image at the bottom of the eyeball turns into a star overhead, a star moreover that does not move, though we and our eyes as we move carry the image with us, and finally that it is the thing, a star, endorsed by our cognition—about all this, the energy scheme has nothing to report. The energy scheme deals with the star as one of the objects observable by us; as to the perceiving of it by the mind, the scheme puts its fingers to its lips and is silent. It may be said to bring us to the threshold of the act of perceiving and there to bid us "good-bye." Its scheme seems to carry up and through the very place and time which correlate with the mental experience, but to do so without one hint further.

The highest order of animal intelligence is the human mind; the highest order of mind is the act of creative intelligence. And this act another neurologist, Sir John Eccles, sees as the most distinguishing of all human activities. Seeking himself to analyze the imaginative process by mapping its movement through the neurons of the brain as they respond to synaptic impulse, Eccles is finally helpless to account for "the illumination [which] often has had the suddenness of a flash, as with Kekulé and the benzene ring, Darwin and the theory of evolution, Hamilton and his equations." But after having ranged so widely through the literature of science, Thorpe gives ultimately the last word to Coleridge: "The primary imagination I hold to be the living power and prime agent of all human perception, and as a repetition in the finite mind of the eternal act of creation."

And so, if *Animal Nature and Human Nature* did not move directly to the point of my question, I had found it nevertheless a book awesome in scope, and the more so because it had given meaning to my own vague wondering, had forced open my perspectives and led to outposts I hadn't known were there—not the outposts traditionally held by humanists, which one could have expected, but new outposts manned by scientists searching for a different and unexpected understanding. And while Thorpe had only hinted at the gathering ecological crises—an inevitable consequence of his concerns—the urgency of his quest was unmistakable: at the edge of such awesome discoveries as anticipated by Ponnamperuma, "It is indeed both a joy and a terror to be living in times such as these—joy at the fullness of life and the opportunities for greater enlightenment; terror at the dangers and disasters threatened." I had found a wise man, I thought, and wondered what he could be like, and wondered now, in the Cambridge pub, whether he could have any idea of quite how far, and in how many directions, his book had moved me.

At 2:00 P.M., I arrive at the designated stairway landing, facing a door of blue-black oak, cracked and seamed with indeterminable age. Because the month is August and the university is not in session, there are no sounds in the building. After a moment's wait, a thin cough from below. And then a black umbrella ascends the stairs, steered from underneath by rubber overshoes—these worn, as the collapsing umbrella reveals, by an elderly man with a gray, bloodless face. Here is W. H. Thorpe.

He whispers a subdued greeting and stoops to work the locks of the ancient door and, behind that one, of a second and more modern one. Behind them both, the cold stone opens to a comfortably vaulted cottage setting: several desks, a couch, various end tables, a piano, two overstuffed armchairs, a Persian rug, bookcases, an electric heater beside which are his slippers—all of it on a Kate Greenaway scale except for the most prominent fixture of all: a giant curved phonograph horn reaching almost to the ceiling, the bell of it easily two feet in diameter. Thorpe removes his overshoes, invites me to take one of the armchairs, and sinks into the couch across from me.

By academic specialty, W. H. Thorpe is an ornithologist. Within his field he is known to be an expert on the vocalization of birds, a study of pure research having little evident applicability to his larger concerns for the traffic of life. By title, he is Emeritus Professor of Animal Ethology at Cambridge. There is available to me no material suggesting anything else about the man.

Certain he will quickly realize my purpose and dispel my confusions, I begin with summary to set the context: when there are too many animals for the carrying capacity of their habitat, as Lamprey and the others have explained, the excess animals die. The human population of four billion is expected to double within the next thirty years. The impending exhaustion of fossil fuels is seen as prelude to the depletion of other natural resources. At the present rate of world use, according to Dasmann, silver, gold, copper, mercury, lead, nickel, tin, and zinc will all be gone by the end of the century. Already amounts of chemical pollution abroad are immeasurable. The consequences of nuclear pollution are unknown. Some of these things are realized now and small fears have been awakened; but others of them are not. The situation amounts to a biological crisis, from what I can see. I cite the concerns voiced by others for the plight of the world rain forests. "They fall now at the rate of fifty acres a minute," I said, "and are expected to be gone by the end of the century."

"Astonishing," Thorpe murmurs, as though much of this were news to him.

I hurry on: the rain forests cleanse the air, affect the weather, maintain water tables. They have been stable for sixty million years. What is it that makes us unable to see the extent of our recklessness against the stores of the geologic past, so that in only a few generations we can use up what has taken millions of years to form? It is his book, I suggest rather heavily now, that provides the background for these questions.

My intention was, after reading his book, to try to go beyond it to report the views of authorities at the heart of their specialties who—

in conversation, talking rather as to a neighbor than to a colleague, talking *beyond* the barriers of specialized language so that all of us might understand them—would simply tell us what is known about our past: where we come from, how we got here, what we ought now to expect of ourselves, what responsibility we owe to the future. . .

He has shrunk deeper into the couch. Peering about the room as though he had misplaced something, he wipes the palm of his hand across his head. He says that he will do his best to answer whatever specific questions I may care to ask but as he says this he seems to be studying something on the other side of the room.

I remind him of a passage in his book discussing the evolution of the brain and the speculations of some neurophysiologists of the warring components within it, the war between reason and emotion. He clasps his hands about his knees and says nothing.

I read him a quotation by a former president of the American Academy of Arts and Sciences, Hudson Hoagland: "Our great cerebral cortex may turn out to be a malignant phylogenetic tumor capable of building incredibly powerful weapons of destruction, but unable to control our hates and fears arising in our primitive limbic brain."

I remind him that one of the neurologists he quotes suggests there exists within the brain "a paranoid streak" and maybe it is this that accounts for human destructiveness. It is said to be a fact, I venture, that man of all the animals wars and destroys within his own species unlike any animal coming before him . . .

In response to this torrent of bad news, Thorpe remains quiet. I realize I have covered my context and had better get now to my principal question.

"If by virtue of our intelligence we have managed to dominate our environment," I said, "why would we be destroying it at obvious expense to ourselves? Is there some basic conflict between animal instinct and human intelligence?"

He clears his throat, unclasps his knees and looks toward the floor. "There's rather slight evidence for what you call instinct in man, except for being born, and so on. What you are getting at is, insofar as instinctive behavior is governed by genetic makeup, then this behavior has been *selected* and therefore it must be consonant with environmental pressures. That is, instincts are there because they are things that work for survival of the species. Intelligence, *too,* of course"—and he pauses here, studying the rug—"but you find that you apply intelligence to every conceivable situation you'll find yourself in.

"The point is, intelligence isn't applied by the race as a *whole.* However intelligent the race is, each little society is applying it for its own immediate needs. What *is* the difficulty is to get the population altogether to come to some sort of agreement where all this exploitation of the environment is leading. See what I mean?"

I'm not sure that I do. I try to steer him back toward biology: "Is it the intention of ethology to suggest a direct relationship in the behavior of animals and man?"

"Some ethologists are concerned with its philosophical implications," he says, "some with applied results. But there has been an increasing interest in the application of ethology to human problems. Hinde, working with apes, has moved now to human problems. And Tinbergen, of course, is interested in the application of it to human problems. The last thing he wrote was on the subject . . ." Later I would look into his reference of Tinbergen, a Nobel laureate whose main studies have been in the behavior of seagulls.

This ethologist sees his science as the rallying ground for a global revolution, and he has confronted his colleagues with the obligation to inform the world public of the biological perils impending. Whatever may be the source of it, there is at present within the human race a fearsome capacity for destructiveness which seems irreversible. In animal terms it is tragically a human dilemma.

"Certainly no one believes human makeup is such that aggression *must* force its way out," Thorpe says now. "But it *is* such that it can be rather easily realized . . ."

In his own book, Thorpe is at pains to distinguish between aggression as the manifestation of an animal's vital need (food, space), and the violence into which aggression can erupt when the animal's need is frustrated. But in the nonhuman animal, violence abates upon resolution of the need, he points out, and very seldom does it lead to death. Collective violence in man—war—is something that is no better understood by biologists than anyone else, nor is its origin clear. (Tinbergen cites the remains of a defensive wall about Jericho, of 8000 B.C., as first evidence of its appearance in the human community.)

"War may have been the end result of the agricultural revolution," Thorpe says. "I don't know . . ." After a long pause, and with some effort, he says: "Among animals there is an example provided by Jane

Goodall of hyenas that war. Tribal wars in which their members are killed. In some cases, ants war between species—a different thing, however. But it is very rare, and the only animals I know of that war are humans, hyenas, and ants." He seems bemused by his own reflection, and he grows silent again.

In his book—where, like Evans, Thorpe is more comfortable with soliloquy—he has something more to say:

> . . . nothing will stop [man's tendency to war] unless mankind as a whole becomes conscious of some superordinate goal for which it must strive with all its strength. Only this perhaps could unite man and bring him to the point where he ceases to bother anymore about war. A gleam of hope comes to me in this—could it be that the terrifying insistence of the increasing pollution problem will be the means of bringing him to his senses?

But this is considerably more than he feels called upon to say now. He clasps his knees and stares at me.

Silence.

So I decide I will just mosey on. Though there is no apparent way to pry conversation from him, however, neither is there any graceful way to end it. "What areas continue to interest you in animal behavior?" I ask.

"The only work I am doing now is in bird vocalization, the specificity of vocal signals and so on—the question of how far there is timbre in a musical sense, and if so what this implies."

"I was very much struck by your writing in your book of the duets of birds. Were they blackbirds?"

"No. African shrikes. They have a very elaborate duetting system." He offers no more.

"How do you record them, Dr. Thorpe?"

"This is done in the field in Africa. It is what I've been about in my recent visits."

"You have annotated these bird calls musically?"

"Yes"—and he goes off to search in his papers for a monograph which now he places before me. "They're found all over Africa; the most interesting ones from my point of view are found in fairly open forests, savanna country, and so on. They're found actually in Nairobi itself. I first had an experimental area on the shore of Lake Nakuru but unfortunately the lakeshore rose. Flooded out. I had to move elsewhere." He stops. I wait. "And then I had an experimental area in northern Kenya, not far from Kitawi. I got absolutely fascinated by this." For the first time, a hint of animation enters his telling. "The normal song of these shrikes is a duet, and it is so precisely timed that people in the Nairobi gardens have heard them for years and never *known* there were two sources for the song. One part from over here, the other from over there, you see? The argument I put forward was that this behavior is not confined to, but is commonest among, tropical birds that live in very dense vegetation, reed beds and so on, and this is the way of keeping in contact with your mate when you can't see. You see? And your mate is the one to give you the right answer to your opening duet. These vocalizations are extremely rapid. The response time is fantastically *quick.*" Interesting. I remember that in his book Thorpe insists there exists within certain species the potential of vocalizing not only for the functional reasons of territory, mating, danger signals, and the like, but for the purpose of creating beauty: an eerie hypothesis. Does he really believe there is an esthetic factor in the formulation of a bird call?

"Yes, um, yes," and he turns back to his papers in search of another monograph he has written; I can look at that if he can find it. He continues to look. He can't find it on his desk, so he moves to a stack of papers on an end table.

"Are you a musicologist, Dr. Thorpe?"

"No, um, I've always been a keen amateur musician, though, and the person I have working with me now is a professional musician." He pauses in his search. "One day she was practicing a Beethoven sonata with the windows wide open and she heard a blackbird, she thought, answering her, you see? She would play a phrase and the blackbird would sing a phrase. She thought the blackbird was changing keys accordingly as *she* changed keys. Having a rather inquiring mind, she immediately went out and got a tape recorder and recorded her playing and the blackbird's response. So then she came to see me and asked if I could tell her whether this was so or not." He has ended his search and returned to the couch. "We had a person in the zoology department then who had absolute pitch, and he came in and listened. The bird did not have absolute pitch."

The silence oozes back into the room.

I have a friend, I say (it is all I have left), a composer who had the experience of a bird perching outside his window as he practiced, and whenever he hit E flat, the bird joined in.

"Yes," he says. "Birds are imitative. They can reproduce whole clusters of sounds. And they can transpose. Yes, yes, of course." He waits. There is no road back.

"What an interesting record machine," I offer inanely, looking across the study to the grotesque apparatus. For the second time, he brightens.

"The best gramophone in the world for 1923. I have an interesting collection of old records, and so I've kept it for the time being. Would you care to hear it?"

"Yes," I murmur. "Yes, indeed."

"You realize what an *enormous* amount of detail there was on the recordings at that time? Couldn't be transmitted by later equipment. I'll put on a Paul Robeson record. Magnificent voice. It was made in London by a firm called E.M.G." He cranks it up, and Paul Robeson sings, to banjo accompaniment, "Way Down Upon the Swanee River."

9

MEMORIES OF HOME

Trout Lake, Washington

Vihtelic faced his second Monday with a special feeling of dread. As he prepared to begin his routine—fishing in water, winding his watch, praying, combing his hair, straightening his clothes, arranging the wares of his inventory, putting far from his mind the frightening dreams of the weekend—he knew that his chances of discovery by a motorist passing above had lessened appreciably. Over the weekend there had been lots of cars, people driving through the park for a day's outing, or climbers traveling as he had from one mountain to the other, sightseers, tourists. He didn't think much any longer about the search party finding him. How long could they be expected to keep looking? He had been down here nine days now, and while he had no doubt that one or several of his brothers were somewhere in the Cascades, still trying, he was pretty certain his best chance lay with the only human contact he had, however remote, and that was with the anonymous drivers passing over the bridge and the gravel road above. But a drive through these woods was hardly a mindless experience! There was plenty to look at, and a road you had to pay attention to (didn't *he* know it), and very little reason to be straining to see, for the desperately few seconds it was possible, to the bottom of the ravine where he lay trapped, flashing his racket-mirror when the sun was out and when it wasn't, just watching them pass by. What he would give for a road flare! If he ever got out of here he would never get in another car without one.

As he threw out his T-shirt again, letting it sink and then slipping it back carefully over the rocks, he decided that Monday, as the days go, was at the bottom of his list. His face was still swollen but so far as he could tell, he had no fever. His wrists and his right ankle weren't swollen, he was still passing urine—it couldn't be kidney failure. Of the things he had to worry about, he decided he would put this at the bottom of his list, too. He still couldn't see behind him, couldn't work today with the tire iron, but whatever his trouble was, it was asymptomatic, and he was plenty grateful for this. His mind seemed clear enough. He would just stick to his routine, which now ordered its own sequences: fish in water ten times an hour, sleep when he could—usually five times an hour—and use the rest of the time to practice with his racket-mirror so that when the sun did come out, he could beam it in at the exact spot the vehicle emerged from the trees. By now he had become so sensitive to such a possibility that he had developed a sort of subliminal awareness he found it difficult to account for. The roar of the stream was there as it had been from the first. But now he could hear anything that was in the slightest way a novel sound, could hear the birds singing over the background sound of the stream, even—and especially—the sound of the combustion engine.

He could hear it—feel it somehow—even when he was asleep! First there was the intimation it was there. He would awaken abruptly, listening hard. Nothing. But he would wait, and in a minute or so, the hushed purr would come, sure enough, rousing him to action. Grabbing for his racket he would take careful aim—dead on target—and be set and ready well before the car or truck arrived. If he was going to get out of here, he guessed now, this would be the way. But of all the days of the week, Monday offered the least hope.

So he worried, and now he began to worry about some things he couldn't do very much about. Mary. Where was she now? At work probably—she would have gone on with things, teaching, holding on to her courage. Mary was tough and Vihtelic was sure she could handle it. What a long time it had taken him to find her, and she had been there all along, in Whitehall, right in his own backyard. Back from the coast, he had dropped into a bar one night with nothing much to do, asked her to dance and soon enough, that was it. They would have been married in June. *Will* be married, he corrected himself.

He heard the purr—no, a heavier sound. It was a truck. He grabbed for his racket, leaned out from the chassis, and aimed the mirror at the opening of the trees. There was no sun. The truck flashed into view and rumbled on, the sound of its raucous diesel fading into the damned singing of the stream. Vihtelic looked for a while at the little fountain of water bubbling up in front of him. Then along the streambed to his left. The birds were back. He whistled to them. He rummaged through his inventory and found his camera. For the hell of it, he leaned out of the chassis as far as he could, aimed it at them and took some pictures.

Mary was okay, but his mother was different. She was the one who was hurting the most, and she was the one who could take it the least. Vihtelic's father had died very suddenly, at the age of fifty-two. It had been hard on all of them, for his children had loved him dearly. He was a strict Catholic and a good man. Catholicism. Even now, in a situation he recognized as dire, even as he had built into his daily schedule his straightforward request to God, Vihtelic put little stock in religion as a value in itself. If his father had been an atheist, which he certainly was not, he still would have been a good man. Whatever the case, his mother was unprepared for the future without him. Somehow she had held all nine of them together, managed somehow to get them all into college. A very big, very close family. Competitive—with seven boys how could you avoid it? But all of them were very sensitive to each other. If one of them fucked up, the older one would climb on him. Charlie had said to him once, "The next time I see you, you better be twenty-five pounds lighter or I'll kick your ass." He lingered over the thought of Charlie. Easily the best athlete of them all. All-conference in high school, scholarship to Northern Michigan. Lifted weights, water-skied all summer, ate a special high-protein diet. A real jock. In his junior year, he tackled a kid so hard the kid didn't get up. The kid was in the hospital for two weeks with internal bleeding. Charlie never played football again. He joined the Peace Corps and went to an atoll in Micronesia, where he lived for two years with thirteen other people. Charlie, Larry, Joe, Frank, James—where *were* they? What were they thinking? What was his mother thinking?

A soft rain was falling now, as it had every afternoon. Water seeped down the side of the chassis again and into the edges of his sleeping bag, creeping through the fabric. The next thing he could look forward to was the sun coming out, and until then, Vihtelic practiced with his racket, shifted his position about the metal lump beneath him, fished in water, slept.

The Miocene

The movements of animals are synchronized to the movements of the earth and the moon in relation to the sun. While this has always been so, the intricacy of the synchrony has only recently and in small part become apparent. It is often difficult to account for. Oysters drop their reproductive cells in the sea but they do so in concert—within a twenty-four-hour period—and never at random. Domestic animals have been observed to grow restive in response to some earth signal of an impending earthquake. Ants, warblers, and lizards journey to distant places by different means, guiding on the sun, or the moon and the stars. Animals react to time and direction through rhythmic coordination with movements of light. Such rhythms take various forms and are common to humans as well, as manifest in the monthly cycles of the female reproductive system.

Emerging fourteen million years ago and moving to its own rhythms, the higher primate called *Ramapithecus* may have been more ape than man, or it may have been more man than ape. From the evidence recovered the line remains blurred. Even now, between modern man and other animals, the line is blurred. While some animals can do things man cannot, other animals variously and by degree can do the things man can do. Dolphins, which must surface every five minutes to breathe, nurse an ill member by pushing it to the surface regularly as required, day and night, until it recovers. In captivity, chimpanzees and a gorilla have responded to human instruction in sign language, and the gorilla, to the satisfaction of some researchers, has demonstrated humanlike cognition:

RESEARCHER *(signing):* Why did you break the bowl?
GORILLA *(signing):* No break bowl.
RESEARCHER: Yes, you did. The bowl is broken. No one else was here. You were alone.
GORILLA: Koko sorry.
RESEARCHER: What does this mean?
GORILLA: Trouble.

The social behavior of animals further blurs the line. In the interest of order within groups, patterns of dominance and submission occur between members and from these arrangements a sense of harmony ensues. This is true of various insects, and of bees, as well as species of mammals, and especially of the higher primates which fashion their lives elaborately to the conventions of the group in which they live. Among these the most successful ground-dwelling animal is the baboon, which lives in groups, or troops, of twenty-five or more. At the center of the baboon troop is the mother–infant relationship (as it is, among humans, at the center of the family, which is at the center of society). For two years the infant is wholly dependent upon the mother, and it remains under her supervision for six more, a time for learning within the community various routines which will favor its survival.

A baboon troop sleeps in trees, or on rocks or cliffs; and it ventures into the savanna during daylight to forage for food, moving out and back as much as twelve miles, arranged in a defensive formation, the strongest members at front and rear to protect the weakest at the center. Routinely vegetarians, they are capable of adapting their needs to the food available. In one recorded instance, a male baboon captured and ate a small gazelle. Soon thereafter other males did the same, and stumbling inadvertently on a method of working together to capture their prey, they adopted their newfound strategy as method. Baboons frighten off their predators by flashing their dagger-like teeth. But when lions appear, they run for the trees. They run from men, too, but remain on the ground hunching low, having learned within the group—the information somehow passed on—that man can shoot them out of trees. Through new opportunities realized, group wisdom forms—the knowledge of all greater than that held by a single member, each member the beneficiary.

By posture, movement, and sound baboons communicate among themselves. Fourteen vocalizations have been recorded, along with thirty-two body gestures, from shoulder-slapping to ear-flattening. They are known to have formed a collaborative system with impalas, which tend to graze in their vicinity. The impala hears well and the baboon sees well. Between the two, each gauging the reactions of the other, danger is kept at a distance. Female baboons remain in the troop for life; male baboons tend to migrate between troops, mating with resident females and thereby reducing conflict and tensions between troops.

Of the movement and ways of living creatures, however imperfectly understood by man (even when he shares aspects of their behavior), the life within the baboon troop, however distantly removed, is believed closer than that of any other animal living today to the life of the new primate, *Ramapithecus,* when it arrived fourteen million years ago.

Gainesville, Florida

Sealed off within my Hertz Fairlane from the heat of Route 295—slipping past flat orange groves, burned-out grass, repetitive billboards promising Lion Kingdom to be drawing ever closer—I shut off the radio and thought about turtles, before Carr. There was Lillian Hellman's snapping turtle, which Dashiell Hammett had caught, decapitated, and left on the back porch, the next day to find it gone—a headless freak, moving on. Hellman had used it as a metaphor of persistence beyond courage, but fixed by her prose to her sensibility, that turtle was a prop. In West Virginia, when I was nine, I bought one at the ten-cent store: the carapace was painted and it didn't last long. In a mud lake I once waded through oleaginous silt to catch a wild one and got leeches on my leg. What else? Phlegmatic, clumsy, aimless: uninteresting. A poor draw for the overflow to Disney World a Turtle Town would be.

But Archie Carr would shrug us all off—tourists, me, and the rest of our indifferent race. Carr sees in turtles things others don't—wondrous things—and, having committed most of his career to trying to understand what he has seen, chasing them all over the world, he is still, at sixty-nine, not much closer to his objective than he was when he began. Nevertheless, he is their brightest student. He is also, as conservationist, their foremost protector; as the author of graceful books about them, their best historian; and as researcher, so zealous in his study of them he has made of their lugubrious ways a science where earlier there had been little more curiosity than the sort I was checking off on my way to see him. And with the green turtle in particular, Carr's zeal escalates to obsession.

The green turtle is a ten- to twenty-million-year-old animal that in modern form weighs about 300 pounds and spends most of its time in the sea. Of marginal interest is the fact that in the water, when pressed, it can swim as fast as a man can run. Beyond these things, however, lumbering onto the land where it ventures only to lay its eggs, it presents to the casual observer a prospect imposing only by virtue of its size.

Well, to begin with, Carr writes, the urgencies of passion heating up his prose, the green turtle (genus *Chelonia*) was an important factor in the colonization of the Americas. On shipdeck, requiring no more space than its size, it could be overturned and kept alive for days, providing food as required. Thereby did the English and Spanish fleets replenish their stores, extend their stay and set down roots. As food, it is both staple and delicacy. The people of the Caribbean have fed from it for centuries; each night, Winston Churchill began dinner with a cup of green turtle soup. Watch a female coming ashore some dark night: ". . . You can see her stop with the backwash foaming flame around her, push her head this way and that with a darting motion less like the slow movement you expect of a green turtle than like a lizard or snake, then lower her head and nose the hard, wet beach as if to smell for telltale signs of generations of ancestors there before her." Sensing at length that *Chelonia* has you in its thrall, he casually observes that fishermen of the Caribbean regard it as having senses "beyond the senses of men," and with this, suddenly he is off in relentless pursuit of its dark mystery, a mystery so profound that Carr's passionate assessment of it almost slips beyond the scientist's control: "It is the most towering of all mysteries; the supreme achievement of evolutionary feats, the ultimate in animal behavior!" But we will come to that later, he says.

A pencil-thin man of no academic airs, informal and easygoing, as Southern as a buttermilk biscuit, Carr wants just a few minutes to finish up in his laboratory—which is, to my untrained and therefore somewhat indifferent eye, a roomful of water tanks holding gaggles of baby turtles—and then he will ride with me in my car to his house where we can talk more comfortably. He has been described to me as a man of extraordinary energy. "He can fly all day to his turtle grounds in Central America," a colleague says, "alight from the plane, and head straight up a mountain without stopping. He will run you ragged."

Do I know where Ascension Island is? Carr asks, climbing into the Fairlane. Well, it's a tiny speck of an island completely isolated in the south Atlantic, about this big—he closes his index finger to within an inch of his thumb. During World War II pilots flying from Recife on the coast of Brazil to the other side had to hit Ascension for refueling or face a thousand miles of open sea before Dakar. "You miss Ascension," they used to sing over beer, "and your wife gets your pension." That's how remote Ascension Island is. Five miles long. Fourteen hundred miles from the coast of Brazil, more than 1,600 from the coast of Africa, on the other side. The way the pilots got there—the ones who got there—was to use their navigational instruments very carefully.

As for the green turtle, a migrant animal, where it lives is not where it breeds. Along the southern coast of Brazil a great aggregation of green turtles gathers to feed on the pasture shoal grasses, departing in two- and four-year cycles for Ascension Island to lay their eggs. (There are no predators on Ascension, whereas there are many along the coast of Brazil.) Hundreds of green turtles arrive at the same time, deposit their eggs—usually a round 100 each—and then turn to the sea and swim back to Brazil. Carr knows this by having tagged some of them.

Now what does it mean that the green turtle can leave the coast of Brazil and swim straight to a five-mile piece of land 1,400 miles away? Assume the animal commences its journey by sighting in on the departing coastline whose features are discernible (*if* it looks back) for no more than fifty miles. Then the coastline disappears from view. And *then* there is the equatorial current flowing in the *opposite* direction against which it must swim some 1,350 more miles. At a traveling rate of twenty to thirty miles a day, perhaps faster, the turtle is susceptible to variable forces—of wind and weather—to the extent that it is more apt to lose its course than hold to it. As it nears the island, a 5,000-foot mountain can be seen on a clear day from fifty miles out. Arriving, say, fifty-three miles off either side of it is—for green turtles or anything else at surface level—pension time: the turtle will arrive, if it ever arrives, somewhere else. Thus, Carr reasons, the Brazilian green turtle, heading for Ascension Island against a strong current and choppy seas, which are characteristic of the south Atlantic, must call upon an open-sea guidance process calibrated within an isosceles triangle over 1,000 miles in altitude and with a base 100 miles wide in order to arrive at its intended destination. It must constantly correct its headings. It must, in short, navigate.

To be certain the indifferent can grasp the enormousness of this achievement, Carr in one of his books has constructed a parable of three island finders all headed for the same place—a sailor in a small boat, a sooty tern (which as it happens flies to the same islands the green turtle swims to), and a turtle. The sailor is the only one of the three permitted to have equipment: a compass, a clock, a sextant, *The Nautical Almanac, The American Practical Navigator,* a set of charts, a parallel rule, a divider, pencils, food, and water. The tern and the turtle come as they are, equipped only with what inheritance and experience have provided them. All three are set adrift and instructed to proceed to their destination.

Though without instruments, the bird and the turtle—to establish their position in relation to the direction they must travel—would have to measure and compare angles made by the sun and the stars with the horizon (as the sailor must with his instruments), and do so with only the naked eye. How they manage this he doesn't know.

> Nevertheless, to keep working toward the general theory let's say that all the animals can somehow measure altitude. . . . But measuring a single altitude is probably not enough. To get a line of position the animal will need to measure, to remember, and to compare at least two altitudes. In the daytime the two altitudes will, of course, be those of the sun, because that is the only body in the sky. Since not much information can be got from the height of the morning or the afternoon sun alone, what has to be done is to take two sightings and extend the arc they make so as to see what the noon altitude of the sun would be. The main trouble here is that, out in the ocean, the sun could not possibly seem to an animal to describe any arc at all, but only to bob up and down. . . .

The bird and the turtle arrive in due course, well ahead of the sailor. Relentlessly Carr works his way through every possibility against known research into animal navigation—even the unverified suggestion that some animals can guide on the stars during daylight—seeking some plausible explanation. There is none. Even more confusing, it is virtually inconceivable that the bird and the turtle would use similar methods. One navigates through the air, the other through water. So there are possibly as many ways to get there as there are animals to make the trip, he concludes, thus adding to his kettle the albatross, warbler, the golden plover, seals, and penguins.

"I don't say the turtle is any better a navigator," Carr says now. "He's probably not as good. I was reading only yesterday that godwits leave the Pribilof Islands and in twenty-five days they come out right on schedule on Piri Atoll. And Piri Atoll is about *this* big"—the index finger is touching the thumb—"smack in the middle of the Pacific

Ocean. I've been there—it seems smaller than Ascension Island, and lower. And the Pacific is a hell of a big ocean." With all its frustrations the search is nevertheless a lyrical one for Carr. In *So Excellent a Fishe,* he writes: "Is the whole Pacific a grid in the genes and brain of the golden plover, do you think? Are the skies of the whole earth mapped by season and time of day in the mind of the arctic tern?"

He has directed me onto a small rural road now where the countryside has moved from the suburbs to assert a sense of place, of pine stands and sand soil, and on we go turning into a smaller sand lane twisting through the woods to his house, which is completely isolated, built low to the ground, air-conditioned cool, and organized about a large glassed-in area that looks out over a freshwater pond. Two parrots are stationed at opposite ends of the room, and lying in the center is a large nondescript dog.

How do they do it? By smell? Taste? Carr doesn't *know.* (A truly maddening question is, how did the *first* one get to Ascension—and *then how did it get back?*) Do they hear the shrimp snapping on the ocean floor and guide off them by some sort of sonar? "If we could just track the damn things! The smell theory, for example. There is a one-knot current coming past Ascension Island. The Koch–Carr–Ehrenfeld hypothesis holds that the turtles are using the smell of the island to home in on it all the way from Brazil. Theoretically, it is possible. We could substantiate it or knock it over if we could track them. When they arrive at the island they're filled with the drive to nest. If you shove them back in the water, they'll come out again. If you take them offshore, they'll come back. If you tie a log onto their front flipper—as people do to mark them for the market—they'll come ashore dragging that log.

"So what you would then do, as you immediately are imagining, would be to go downstream two hundred miles and release some turtles, and then go across the current and release the same number. If the ones you release downstream show conspicuously more success in maintaining course toward Ascension, then you're strongly suggesting that olfaction, or some taste–smell emanation from the island, is involved. But first you've got to track the damn things!"

To this end, he says, and for many moons, he has aspired. The turtles are willing. Their shells are so thick and lifeless about the edges that perforations for securing lines or holding devices can be made without damage or pain to the animals. So long as they are still able to move, they will haul after them anything attached to them.

At first he tried floats painted orange and attached by a twenty-foot line, designed to trail after the turtle when it dived. Maximum range: one mile. Then he tried air-filled balloons painted red and tied to the float. The weather blew them down. Then he tried weather balloons filled with helium and painted orange; he picked up a few miles more. Another zoologist who was trying to harness radio transmitters to whales for the same purpose gave up and Carr inherited his transmitters. He attached one to the back of a turtle but the signal was too low to the water for successful transmission. By now he had some blimp-shaped balloons that were more stable even than the weather balloons in high winds, extending his range to nine miles. He attached a transmitter to a blimp balloon which was attached to a float attached to the turtle. Maximum range: nine miles.

In 1967, with the cultivated patience of one who, resigned to frustration, has developed the steel control necessary to contain it, Carr wrote: "The ideal way to maintain contact with island-seeking turtles cruising in the open sea would be by satellite."

"Then NASA came to *me,*" he says now, still surprised. "They were in their ebullient mood of thinking, hell, we'll get a lot of credit with the world if we help zoologists learn stuff about animals—help out in measuring the earth's assets and all that."

"But NASA is a space agency. Why would it want credit for zoology?"

"Well, there's always been some question of whether the space program is realistically necessary for the welfare of mankind. A lot of people say, why don't we improve the state of man on earth rather than all this business of going to the moon?" (Howard Ensign Evans, I remember, for one.) "So one thing NASA tried to do was show they were polyvalent in their attitudes, that they could help all kinds of people. And they offered to help me." There was a spinoff possibility, too. If Carr could figure out how turtles navigate, the information might be adapted to submarines, missiles, and other nonbiological travelers designed to protect men from each other.

NASA and Carr made their plans for tracking green turtles, wired to more sophisticated transmitters, across the sea. "I had the address on the satellite and they were going to shove the plot graph out of the computer in Long Island and send me back here a prepared file of the movements of that turtle, together with any physiological parameters

I wanted to measure along the way. And the same with the polar bear and the elk, and the whale."

"The polar bear, the elk, and the whale?" I ask, wondering where this odyssey will end. One of the parrots shrieks. Lost in his account, Carr ignores us both.

"They gave me an engineer, and I got a grant from the navy to do the work and went out myself to Ascension Island. We worked at it off and on for eight years. The turtle would carry the apparatus down under the water, and it would leak, or the turtle would break the antenna off. Or God knows what. I hate electronics. They scare me now and they always have." (He taught physics to air cadets during the war, I later learn.) "I never *yet* have got hold of an engineer who could lick both the electronics problem and the sealing problem. All this just about turned me away from turtles. . . ." He asks me if I want a beer; I say sure. When he comes back from the refrigerator his mood is unchanged.

"The most nerve-racking experience I've ever had in my life was to see this *déjà vu* thing—we'd get a *new* engineer, and he'd come with his packing incorrect and then the water would seep in. A nightmare! You'd watch those guys tinkering around, spending money like water —seven thousand dollars a float!—sending it out to sea with the turtle, watch it sputter and smoke. After eight years of that, NASA, who had been farming out such projects to other zoologists—a polar bear one, an elk one, a whale one—"

"A polar bear?"

"Then they had an *ape* they'd put in a satellite. It came down near Honolulu and it was all right for a little while but it died. The whale stranded itself and died. The polar bear got sick and somehow they caught it with a net and took the thing off. And the elk died of pneumonia. NASA was embarrassed horribly. Four reverses involving animals. Fortunately no turtle died. They're nervous as hell over their image. They don't want you to do *anything* that would cause the public to think they're encouraging cruelty to animals. I think they made too big a deal out of it myself. Anyway, I just decided that until their people had perfected marine-going radio equipment that could be installed in floats, could be dragged underwater two hundred feet, and would last for several months, then I wouldn't try to continue. But I don't mean to say that my confidence in this kind of experiment has dimmed. I was telling myself just this morning the single most important thing anybody can do today is to solve this most towering of mysteries, animal navigation across open water to a point goal." Morosely he sips from his beer. "All you know now is where they leave from, and where they show up."

There is a minor footnote, I remember now, to the discovery by Penzias and Wilson of the background radiation flowing through the universe, the cool echo of creation. One day a pair of homing pigeons were discovered nesting in the bell of their listening device. The scientists removed the birds carefully and transported them fifty miles away to release them. The next morning they were back. Maybe Carr is right in his extravagant evaluations of such things—is it knowledge of more importance even than the time measurements of space? Certainly Carr's parable forces more vividly the sort of question I had tried to put before Thorpe: what is the nature of the distance between the intelligence of the sailor and the instinct of the bird and the turtle? I tell him of the lengths Thorpe has gone to in his book to narrow this distance, and of his hesitancy to deal with it when I saw him.

"So much is known now about animal sociology," Carr says. "Anybody who doubts there is extremely complicated and ecologically advantageous social organization among animals is off his rocker. And what the difference is, I fail to see—none whatsoever between the social organization of wolves and that of man. Except that of wolves is a lot more successful!"

"How are wolves more successfully organized than people?"

"They don't kill each other! They don't have wars. Of course they don't relieve pain or do any of those humanistic things that we do, but *biologically* theirs is in the long run a more successful social organization than what we have—which seems sure eventually to obliterate us from the face of the earth." He adds this last as though he doesn't expect me to take him up on it—and I don't, for the moment. I take him back to a similarity—to the new language experiments with apes.

"I don't know how you amass enough critical data so that you can ever explain exactly how an animal puts across ideas to the degree that I believe they *are* putting across ideas." He looks hard at me. "Because I believe they are *all the time!* And it's the experimental approach that puzzles me. Trying to teach animals human words isn't fair. It's interesting, but I don't think it has much to do with the real problem, which is, how do a pair of pointer dogs seem to be constantly in contact with each other when they're quartering a quail field?

What's the *degree* of communication? What kind is it? That's a form of language as I see it.

"That parrot over there"—he points to the one behind him—"has a beautiful form of language. Not human speech, although it knows thirty-two words, mostly obscene and Spanish. But that's not what I mean. It calls water 'tinkle-dinkle-dee,' which is onomatopoetic, the sound that comes when people jiggle her dish when changing her water. A little bubbling sound. This parrot"—he points to the one behind me—"announces any hawk out front with t-h-r-r-u-p. Or even a big heron, or even the shadow of its wings. T-h-r-r-r-u-p. I think the sound comes straight from the rain forest. It's what parrots must say to each other when something shocking is going on, say, when a harpy eagle is about to descend. Well, the parrot behind you began to attach that sound to any strange arrival out front. A person, for example. And I think *that* brought about an interchange between her and the *dog*. Because now, when she goes t-h-r-r-u-p, the dog springs to his feet and starts bellowing. And he will bellow whether it's a person coming in or just a sparrow hawk flying down there over the lake. So *he* has come to pick up the same thing we did. It took him a lot longer to do it, but he realized that in parrot language, t-h-r-r-r-u-p means, look out, boys, something bad's coming. And now, if *he* starts barking out there, *she* will say t-h-r-r-r-u-p." At the moment, except for an occasional screech of unknown intent, the three of them ignore us.

For Carr, then, distinctions are more intriguing than the similarities. "Sometimes to a zoologist," he goes on, "it is really almost incredible in the degree to which intelligent, nonscientific people get confused with respect to, well, sea-turtle navigation and the navigation of a man flying a plane across the Pacific. It's the same ecological accomplishment, but the mechanism—the nervous equipment and the biological equipment involved—are *totally* different. It's an *apartness,*" he says, visibly working for the distinction, "an apartness of instinctive patterns as compared with this other way of handling the exigencies of ecology, which is to be, as *we* call it, generalized to the extent that we can react to any new situation and devise our own patterns. But our plans are often improper, too.

"So we can't *do* what animals do. Because if you're going to rely on generalizations, and a generalized superior nervous system with a hell of a lot more coordinating centers in it to handle every problem, need, or necessity that comes up—every situation in which you have to manipulate objects—then feats such as these have to be *learned.* The instinctive pattern has the great beauty in it that it doesn't have to be learned. It may have to be *triggered,* or imprinted, but it doesn't have to be learned all over again."

"Have you thought of what point it was that man began to lose this?"

"Yes. Where you want to look is in the bushmen, in the Australian Aborigine. The way those guys find water seems to me to show that they have *not* lost it. I really believe there is some sort of an instinctive turning of Kalahari or Australian Aborigine toward water. I haven't the foggiest idea of how they do it, but they have the ability to find where it's raining even though you can't see any difference in the weather all the way to the horizon! They'll walk sometimes up to fifty miles, totally attuned to meteorological conditions." Carr once experienced those talents firsthand on a month-long hike through the Nicaraguan rain forest. "By God, those Indians! The expression itself violates our sense of science, but they had a *sense of direction!* You could get in a howling rain forest where everything was exactly the same, as much as on the bottom of the sea, and in this absolutely amorphous environment, with the trees closing a hundred and twenty feet above your head, and every tree looking like every other tree, those guys would sit on the floor of the forest and argue out the direction to the camp we had left four days ago!"

Once, on the island of Yap in Micronesia, I had heard of similar skills attributed to local fishermen. As boys, they were said to lie on their backs in the water to "sense" the ocean currents, and they navigated open sea, sometimes hundred of miles, by splashing water into the boat and studying its motion within the bow. In Kenya, an animal ecologist out of Oxford showed me a road cut by an African working without surveyors or assistants, beginning at the top of a hill and bulldozing over flat bushland straight to the horizon, fifteen miles away. Measured later by instruments, the line was true within inches, from beginning to end.

I told Carr the others' emphasis on the beginning of agriculture as the turning point in human history. Could this have marked as well the beginning of the loss of reliance on instinct?

"I think they are absolutely right in the time periods suggested, but whether one was the cause and the other the effect, I don't know. The word *intelligence*—whatever *that* means; I don't use the word much because I think our dog is pretty intelligent—the one feature of it I've never been able to see in an animal is prescience, or anxiety, things that depend on your knowing there is a future. And I was always told that animals don't sit around dreading their own death or being attacked by enemies, because they can't think into future time.

"Well, if you can't think through a future time, you're not going to be a farmer because you've got to plant a seed. And the thought that that seed will grow can come to you only by two ways. One is, instinct tells you that's the thing to do and you survive by doing it although you never know *why* you do it. The other way is, you have achieved a certain level where prescience has come into you, and you know that seeds when put into the ground sprout and grow into a plant. So I don't *know* whether farming got started only after man developed a certain ability to look forward to and guard himself against enemies by building walls and tree houses and abandoning whatever equipment the apes had for doing exactly the same job on an instinctive basis. What I'm saying is, I don't know what intelligence is. But what people seem to come back to, after they get over tool-usage and language, is anxiety over the future. Any form of awareness really. Now I can't be sure that my dog doesn't feel all those things in a subdued way."

Nor can I, but I am almost too embarrassed to speak of it. My own dog was a Vizsla, and as I struggled with the implications of these conversations and how to put them forward, he was for a time my only companion. On one bitterly cold night he manifested such awareness I was startled by it. Since I had occasionally to travel a hundred miles from my house in the woods of upstate New York into the city, and often on short notice, I made arrangements with a family named Duff to keep him while I was gone. To ready him for this situation, he was taken in the car to the Duffs' and left overnight. The drive was up a sharp mountain slope and through winding woods, and down the other side of the ridge, a distance of two miles. My dog hated the cold —Vizslas are a short-hair breed—but if I left him in the house overnight he began to bark at dawn to be let in the bedroom. On that night I sent him out of my house to his at about eleven. During the night the temperature fell to fifteen degrees below zero. Around four I was awakened by his barking—I thought sleepily at the time a car had pulled into the drive and this had set him off. But there were no lights outside so I went back to sleep. The next morning the dog was gone. About nine, Mrs. Duff called and said the dog had come to her house at four and barked—she said, and I'm sure she was right—to be let inside. Two miles over ground it had never run, acting on some impulse of selection, choice, and possibility.

But I keep quiet about it. Carr is the expert; let him find reasons for such things. "No animal is aware of death," he is saying, "nor does any animal, they *say,* sit around dreading dangers until the smell or sound of some sign of the danger comes in. Do you believe that? I don't know if it's true or not."

"But we still seem to have instinctive physical fears—of height or water. Why aren't we frightened by the threats to ourselves from our own actions in destroying our own habitat?"

"I've given that some thought" Carr says, frowning. "But it gets me confused. The conflict between the forces which tend to make us abandon our animal ways and the advantages of being animal! One of the animal's advantages is to have ready-made, instinctive, successful behavior patterns to call up. If *we* had them, we would never, never engage in war. But by becoming human, we get all these other attributes—avarice, jealousy, ambition—and along with these we get the feeling that we can cope with our own society. We believe we don't need to rely on instinct, whereas this just has proved *not* to be true. We've created a monster here—our own social-political-economic organization that we are obviously *not* capable of handling."

"How?"

"How did we create it? I don't know. Our intelligence has enabled us to conquer the forest, and we've put aside the ax, and we don't have to walk around with guns to kill food very much anymore, but I think now that we have built such a complex civilization that we lack the intelligence to run it. In other words, we lack the intelligence to run the situation that our intelligence has built! And I don't think this is a paradox at all. I think it was more or less inevitable."

Does he mean that intelligence is inherently destructive?

"Not in any purposeful fashion," Carr says, "but as a force, intelligence is *inevitably* going to have itself in too complex a system to manage. Maybe the dilemma it is causing us now could have been avoided, and maybe it can possibly be circumvented, but I think it's fair to say that we lack the intelligence to cope with the situation here on this planet that our intelligence has put together."

"Would you say then that the exchange of instinct for intelligence has caused these problems?"

"Not exactly. There's a limit to what genetics can do for you in terms of ready-made instinctive patterns to cope with all types of civilization. The potter wasp, being born with the behavior pattern to build a beautiful little jug out of clay, is one thing. But for man to be born with any resource that he can suddenly call forth to cope with the complexity of factors in the chaos he has wrought in the world —that's just expecting too much of genetics. But it's not just the brain either. In the short time we've been here, evolution could never have evolved us in ways to be able to handle these rapidly accelerating complexities. Polluting the atmosphere, the continental waters of the United States! I don't like to keep going back to that word intelligence, but to me it's an indictment of human intelligence that everybody *isn't* aware of these problems and in a high state of excitement to do something about them."

10

LETTERS TO MARY

Trout Lake, Washington

Vihtelic found a piece of wrapping paper in the glove compartment, and of course he already had a pencil. On Sunday he wrote:

September 19, 1976
8 o'clock P.M., E.S.T.

Mary,
This has been a most difficult day for me today. I have been waiting and expecting to talk to you all day and tell you how much I love you and care for you and that I'm all right. But nothing has worked out for us. I was supposed to have been removed from the car—my left foot is still pinned—and transported to a hospital 50 miles from here. They never showed and never showed.

John

On Monday, he wrote:

September 20, 1976

My Love,
The love I feel for you now, Mary, is so intense and hot that I have to start writing again even though the pain is bad. One thing I realize now is that all of those people that were to come and get me were just dreams and no

one really knows where I am including myself. One thing I do know for sure is how much I love you and how good I will be to you if you will have me in my deformed state. I really believe now that the foot is dead and will have to be removed.

John

On Tuesday, Vihtelic wrote:

September 21, 1976

Mary,
The reason why I write so little to you each time is because it is such a hard position to get into, and I have to cut on my left foot that's help me walk either. This really is a peaceful place. I'm lying in the middle of a washout. I'm sure in spring there is water on this side too. It's so hard for me to write to you without crying I think I'll stop for a while.

John

On Wednesday, he wrote:

September 22, 1976
12 P.M., P.S.T.

Love,
It's really difficult for me to believe that I have been out here for eleven days. So many things amaze me right now I don't know where to begin. The fact that I am here in the first place. The fact that no one looked over that ledge since I've been here is I don't know what! A lot of the time I think of what you might be doing at that moment. It's really very sad but this is a sad time for many of us.

John

The Pliocene

Less is known by man about the origins of himself than of most other animals. If *Ramapithecus* was the first hint of us, what did it look like? Could it think and talk, laugh? Could it know it would die? How did it spend its days? What became of it? To the first hint of man there are only hints at answers.

Ramapithecus was smaller than an ape but larger than a monkey. It had long arms like an ape, but its face was smaller than the ape's, and its nose, or snout—in relation to the rest of its features—was diminished by comparison. Its feet were suited for climbing trees, which is where it lived during the night, climbing down in the daylight hours to forage the plains for food. As a higher primate, as high at least as the ape, it doubtless shared the behavior known now to characterize modern primates below the order of man. That is, it taught and learned, thought and planned.

What is known of *Ramapithecus* that is closest to fact is based on the fossil remains of thirty individuals, most of which lived between fourteen and ten million years ago. They were found in areas which once were forests edging onto plains—in Hungary, Turkey, Kenya, India, China, and elsewhere. In each instance, the most that was found consisted of jawbones and teeth. (Jawbones, because of their mechanical strength, and teeth, because of their hard enamel, last longer than anything else.) It is little enough to go on but more can be surmised than might be expected because of the striking variance of one of those teeth. All the fossil teeth found from the remains of *Ramapithecus* are the teeth of an ape except for the canine tooth, which is the tooth of a man.

But even in their modern form, and allowing for what is evident enough to the eye, the biological differences between modern men and apes are very small. Of the several hundred amino acids in the hemoglobin of man as compared with the gorilla, only one is different. In the measurements of molecular taxonomy—a system of biochemical analysis comparing immunological reactions with protein and nucleotide sequence in DNA to establish classification—the differences are minute. By such a scale as this, if a value of one is given to the difference between African and South American monkeys, the difference between man and the chimpanzee is only .12, or one-eighth of one unit. This is closer than the fox is to the dog; about the same as the difference between a zebra and a horse.

So even though it is only a single tooth, the canine of *Ramapithecus,* more man's than ape's, is an important distinction from which much has been surmised. In the ape the canine is extended and sharp-pointed, daggerlike, and used to strip bark from trees; when bared against predators, it is a signal of threat. The canine of the ape is pickax and sword. But the canine of *Ramapithecus* is small and rounded, a tooth built more for chewing than slashing and tearing. As its earliest ancestor had moved into the trees to change its diet, *Ramapithecus* (so its canine tooth suggests) had moved out of the trees in search of different foods, seeds, and grass. In the savanna, exposed and vulnerable, it may have organized its kind into troops similar to those of the baboon, rising often to its rear legs because it had to do so in search of opportunity, or to watch out for danger. And if it did rise its hands would eventually have become free for other things.

From ten million years ago through the next six, there is very little evidence of *Ramapithecus:* isolated teeth at nine and seven million years, a lower jaw with one tooth in it at five and a half million years —and then nothing at all. What happened to it through this long dark period of its evolution—an interval 600 times longer than modern man has lived—is unknown.

Nairobi, Kenya

"Right there, in the canine tooth of *Ramapithecus,*" Leakey is saying, "is the first bone of contention." He smiles wanly, as do I—two self-aware members of *Homo sapiens sapiens* in the dining room of the Norfolk Hotel twelve million years or so later, lunching over poached fish and fruit, comfortably enough except that the eyes of this one are still watering from that head cold. I switch out for a moment to wonder what it will do to my schedule if I take another day here, to shake it.

The waiter says something to me which I cannot understand. "He asks if you are finished," Leakey says, and when I respond that I am, apologizing for my clogged ears, he turns to the waiter and informs him so in Swahili, somewhat more courteously than most elsewhere would take the time to do, whatever the language. The domino effect: if I delay here, what will happen farther on? It is impossible to change the schedule. As compensation, my sinus for the first time in two days allows a breath of air to slip through a channel somewhere in the upper right quadrant of my head.

"But in the case of *Ramapithecus,* the teeth are not just a first measurement," Leakey is saying, "they are the *only* measurement. Now it has been argued that when we lost the canine we had to develop aggressive tendencies because we'd lost all else. I think a more plausible argument is that by dealing with the new food sources now available in open country you needed a different type of chewing apparatus—instead of chomping up and down, a more rotational movement . . ."

Richard Leakey as a boy once assisted his father in demonstrating that early man could compete with larger mammals in finding meat for himself. Nude and wielding giraffe bones along the edge of the

Serengeti Plain, the two of them drove a hyena pack away from a zebra carcass and tore off the remains of a leg. His mother, Mary, is at this moment studying footprints left in lava ash to the southwest of here more than three million years ago. His father, the late Louis S. B. Leakey, whose discoveries of stone tools and ancient bones outline much of the knowledge currently available concerning human evolution, uncovered some years ago the first fragment of the earliest ancestor of *Ramapithecus,* dated to twenty million years ago. This Leakey, Richard Erskine Frere Leakey, is the great-great-great-great-grandson of John Frere, the first man in history to record the discovery of prehistoric flint tools; in 1797 he reported to the Royal Society of London his theory they had been made by creatures unknown, some people evidently coming before Adam. At thirty-four, a paleontologist by nature and by nurture, although without formal university training, Richard Leakey is regarded by many as the most accomplished yet of his illustrious line.

Ceiling fans stir the dry air. The Norfolk is among the last of Nairobi's colonial settings. It is cool and quiet in here. "The story is delightfully vague," Leakey is saying now, "which I think is essential to stress at the beginning because nobody knows what happened." He fingers his unlit pipe. "But the scenario that is *imagined* is that sometime during the latter stages of the Miocene—thirteen to ten million years ago—there appeared in Africa and other parts of the tropics the beginnings of grassland adaptations in the animal groups —antelope, the gazelles, various grazers. And this coincided with the forest shrinking and the grasslands opening up, presumably a reflection of a climatological change. And then you had *Ramapithecus.*"

"Is there anything that precedes *Ramapithecus?*"

"It is the earliest," he says. "There's nothing about man until you get up here, about fourteen million years ago."

"Could it walk upright?"

"We don't know. There are no limb bones extant, and you can't tell about stance from teeth. Between ten and four million years ago there are only a few specimens, but by the time you get to four million, the story is much better documented. By three and a half million, there are specimens from both Ethiopia and Tanzania, which consist not only of jaws and teeth but parts of skulls and limb bones. And in Ethiopia there is the specimen Don Johanson collected which he has named Lucy."

Although it is recognized as one of the most significant fossil discoveries ever, Lucy, now three years after emerging from the ground, eludes definitive interpretation. Her name derives from "Lucy in the Sky with Diamonds," the selection playing on the tape deck in the camp when Johanson, an American paleontologist, and his crew returned with news of their discovery of the most complete fossil of a prehuman ever found. From the neck up, with the important exception of her jaws and teeth, she was more ape than man. But among her remains the pelvic structure survives as well, and from this it is unmistakably clear that Lucy could walk.

From walking—standing and moving upright, from what essentially was nothing more than a change in posture—came humanness. It seems an odd way for evolution to have introduced us. One of the most cautious and conservative of researchers is Richard's mother, Mary. So important is this single feature, she recently wrote, that

> one cannot overemphasize the role of bipedalism in hominid development. It stands as perhaps the salient point that differentiated the forebears of man from other primates. This unique ability freed the hands for myriad possibilities—carrying, tool-making, intricate manipulation. From this single development, in fact, stems all modern technology. Somewhat oversimplified, the formula holds that this new freedom of limbs posed a challenge. The brain expanded to meet it. And mankind was formed.

Hugh Lamprey's insistence on the cumbersome term "biofeedback" to explain such incredible developments is surely appropriate here: in size and complexity the brain grew because of a change in locomotion! Challenge and response. Whenever it first occurred, whether for the first time with Lucy's people 3.6 million years ago, or earlier, standing up and walking changed the animal which changed the environment and, in time, changed the world.

Reflecting on this epochal event within a context somewhat removed from Mary Leakey's, Sigmund Freud believed it changed the nature of pre-man's sexuality, too. At first there was a shift in the functions of the senses, the visual taking primacy over the olfactory. (It is now known that the olfactory nerve cells in the human embryo begin to die before birth.) By standing up, Freud wrote, the male shifted his attention from the cyclic odor of estrus in the female to the always accessible view of her genitalia. "Sexual excitation became constant and the family was formed, and so to the threshold of human culture." A less exalted but no less searching reaction to this transformation has been expressed by the biologist Stephen Gould who, writing of these developments as well as Freud's version of them, declared in his last paragraph:

> I'm finished. I think I'll walk over to the refrigerator and get a beer; then I'll go to sleep. Culture-bound creature that I am, the dream I will have in an hour or so when I'm supine astounds me ever so much more than the stroll I will now perform perpendicular to the floor.

But if *Ramapithecus* came first, who then was *this* walking creature? "My impression is that Lucy may well be a later form of *Ramapithecus,*" Leakey says, finally firing up his pipe. "A small creature, extremely versatile. She'd retained a lot of the climbing adaptations, so she was also agile in the trees. While she was capable of upright stance, this hadn't developed to the point that it has in man."

If Leakey is right, the arrival of man comes after a very long voyage, the slow descent from *Ramapithecus* of fourteen million years ago to the three-and-a-half-million-year-old Lucy, its direct descendant. But in Leakey's version, other hominids as well would have separated from the *Ramapithecus* line much earlier, with the consequence that Lucy in her moribund state was still around at the same time as others who would eventually outlast her. (The creatures who left the footprints Mary Leakey's team discovered in southwest Tanzania are believed to be contemporaries of Lucy.) We ourselves are the only species surviving those times, but existing then were two varieties of sluggish cousins called australopithecines (southern apes), both more ape than man. Paleontology distinguishes between them by their dentition—one a big-toothed, the other a small-toothed species. And there was as well, or possibly coming soon thereafter, the first line of the genus *Homo,* our most immediate ancestor. Down *Homo*'s line would evolve *Homo habilis,* or tool-making man, to *Homo erectus,* who added fire-keeping and hunting to his skills and, by one and a half million years ago, had achieved the classification, without hedging, of man.

But the man who discovered Lucy, Don Johanson, disagrees with Richard Leakey. Lucy, Johanson claims, is a separate and distinctive species, not a *Ramapithecus,* but an australopithecine, and as such, the mother of hominids—of subsequent australopithecines and the *Homo* line; in effect, she is not very far down the line from the point at which man and ape separated from common stock. This occurred more recently than Leakey supposes—perhaps no later than five or six million years ago. Molecular analysis, which is used now to date the divergence of various species, tends to support this argument.

"But that's a fairly circular argument," Leakey says. "What we have found to be certain is that *Australopithecus* has been discovered in East Africa and South Africa—the only two places that genus is known. In my opinion *Australopithecus* does not go back as far as some of these early specimens," which would include the findings of his mother as well. "So they are either another species of *Australopithecus* entirely, which is an academic decision, or they may well represent an early species of the genus *Homo*—which is also an academic decision. One would have to weigh the anatomical evidence to decide which way to put it," Leakey says. "Whether you want to qualify *Homo* on the basis of the brain; or whether on the basis of the bipedal stance and the development of the body."

But the issue is surely important beyond academic definition, for it speaks to the question of time: *when* did man in his earliest form first emerge? Leakey is embattled on the point, however, and there is no value in pressing it.

"If you were trying to trace the emergence of man through intelligence," I asked him, "how would you characterize the intelligence of *Australopithecus?*"

"That's a question I'd rather you ask other people," he said, "because I don't know what intelligence is." An echo of Carr! "A very complicated question, a program on its own. What do we mean by it? The ability to conceptualize? To make complicated sounds that take the form of speech as opposed to a series of sounds that are not in any way coordinated? I think australopithecines were less sophisticated in their interactions, in their social codes, in their ability to coordinate and cooperate in gathering food, in their ability to pass on information to one another. Could australopithecines make tools? There is an assumption that they could. My position is they may have used natural objects in the way a chimpanzee does—modified branches and sticks, maybe, to pry at things. But the deliberate process of taking a couple of stones and using one to manufacture from another an object of predetermined shape that could be used for a predetermined function—I don't think this was possible for *Australopithecus.* And there is no evidence to suggest it was. Now, how you would distinguish the potential intelligence of *Homo* and *Australopithecus,* I don't know. I think Ralph Holloway would have a lot of ideas on this. You should ask him."

I did, later, at Columbia University in New York, where he is in residence. Holloway's specialty is the comparative analysis of prehistoric man's brain through casting plastic molds of fossil skulls and deriving from these the external shape of the brain itself. Like Thorpe, Carr, and now Leakey, too, Holloway says he doesn't know what intelligence is, but he believes it has to do with the *organization* of the brain, not so much with its relative size, nor with its subsequent evolutionary growth in later *Homo* species. In some humans with abnormally small brains—no larger than the ape's, which is about a third the size of modern man's—there is still the facility for speech, human reasoning, and so on.

In any event, Leakey goes on, about a million and a half years ago the line of man's descent finally became clearly discernible. This was with the arrival of *Homo erectus,* the first animal to be classified as man. "And we now have the material of about half a million to about three hundred thousand which shows the transition from *erectus* to *sapiens.* And so you have this beautiful story of the continuum with the genus *Homo,* from present day, back through the fossil record—back to two and a half million years ago, and *I* think three and a half million. But that last million years is under dispute at the moment as to whether you use the term *Homo* or not—because of the brain."

"What is that dispute about?"

"The first question is, is absolute size of the brain significant? The other area of controversy has to do with brain volume."

Before leaving New York I had read somewhere that the size of the brain of man had *tripled* within the past 400,000 years. I asked Leakey, "If the brain is now three times the size it was in *Homo habilis,* wouldn't this in any case be a major factor in the evolution of modern man?"

"Well, sure it's been used as a criterion," he says, "but what is significant is the development of that volume. The question is, should you think of *Homo* as a biological adaptation toward intelligence? Is that what distinguishes us from other primates—our greater ability to think, to plan, to do? Most people would say yes, since it's in this area of intelligence, or behavior, that we have gone furthest. We've become extremely sophisticated in the coordination of hand and brain, coordination of tongue and brain—of speech. Does this result from a large brain, or is it a development of the brain irrespective of size? People have argued convincingly that the average person today uses less than 30 percent of his brain. If you find a creature with seven hundred ccs of brain volume, and your apes have five hundred ccs, and *Australopithecus* had five hundred ccs, do you say that because a brain is seven hundred ccs it should be classified as *Homo?* Or do you say it's the shape of the brain and the development of certain areas of the brain that determine whether it's *Homo* or not? Is it the development of the frontal lobes, the temporal lobes, the development of the speech area?"

Within his choice of ways to think of the development of *Homo,* what does he see as the significance of the longer period of dependency in the human infant?

"It's significant in two ways," Leakey says. "First of all, you have a longer period for learning, which no doubt is important. But you also have a longer period of vulnerability—you are defenseless. This is in some ways an unfortunate adaptive situation. You're subject to predation, disease. A species unable to look after itself in the infant state is an unsatisfactory species. This is the big weakness in man, and in most primates—but particularly in man. Was this delayed infancy a requirement for *Homo* in that it needed a long period of infancy just to mature, or was it necessary for the infant—to get through the birth canal—to have a small head that would grow after birth? My own impression is that the birth of a large-headed infant is a functional problem, an anatomical difficulty. I think the problem came first."

"But if adaptation is a response to problems, what problem would have brought this about?"

"Presumably, with the development of the brain, with the greater dependence on the selection for intelligence, the process was helped along by the increasing brain size. And with any increase in brain size, you get the consequential problems of birth. And you reach a point where you can't get your birth canal any larger. Because if you widen your hips and you're still going to walk upright, there will come a point when you're no longer able to support the weight. So if you can't widen your hips anymore, the alternative is to produce an infant that will grow more slowly."

"Is it known when language emerged?"

"No. But I would take the position that language is a prerequisite for the sort of life that goes with technology. If you're going to make stone tools, and if you're going to come together to live in campsites and share labor responsibilities—all of which I think is the package that goes with mankind—I would have thought some form of language was essential, that it emerged at about the same time as you're seeing these other things emerge." Holloway believes there is an even stronger inference to be drawn in tracing the early existence of language from the evidence of common characteristics found in certain early tools. How could they be made by a variety of individuals without the tool-maker having passed on instruction? The evolution of tools corresponds to the evolution of humanness.

"The significance of structural apparatus in the brain is still debatable," Leakey goes on. "There's no question that *Homo habilis* of two million years ago had the structure for language, as opposed to a gorilla or chimpanzee. Now this doesn't mean that he spoke, only that the *possibilities* that he did are stronger."

Is the ability to use language then what Leakey thinks separates man from beast?

"I don't think you can separate man from beast because we *are* beasts. We are part and parcel of the same thing. If there is a difference it is our extension of ourselves in behavioral ways—in the ability to manipulate our hands, and in the extraordinarily complicated development of the brain to provide for speech and complex communication. But it's not the brain alone! It's the combination of the evolution of the brain and the body. I think our difference is part of a package, not a component."

"Is there any unanimity among anthropologists about the beginnings of civilization?" I asked him. "Did it begin with agriculture?"

"Agriculture is part of technology."

"Then when did technology begin?"

"Two and a half million years ago, in East Africa. With tools made of stone by *Homo habilis.* Technology has a very long history."

"How did it evolve?"

"There were just more tools, but increasingly complex ones. But of course there's a limit as to how complex a piece of stone can become. So what you see is a series of different-shaped stone implements made for different purposes."

But in his book with Roger Lewin, *Origins,* Leakey is less inclined to subsume agriculture so quickly:

> There can be no doubt that the sharpest and most dramatic shift of gear in our ancestors' progress along the path of human evolution was the invention of agriculture ten thousand years ago. That shift, from an essen-

tially mobile hunting and gathering existence to an essentially sedentary agricultural economy, shattered a way of life that had at first emerged at least three million years earlier, and was responsible for creating the basics of humanity in the way they are. The invention of agriculture was, without exaggeration, the most significant event in the history of mankind.

The table has been cleared, and I notice now that we are the only two customers left in the room. Leakey has other appointments for the afternoon but he seems willing to stay with it until I finish. I order more coffee. But if it was not the advent of agriculture that marked the advent of civilization, I ask him, why would agriculture have taken so long to occur? As has by now become evident, Leakey favors rhetorical questions over direct answers, and he responds in a short burst with two of them: "What do we mean by civilization? What is civilization a product of ?" I pour us coffee and wait. "As I see it, it is a stage in history when humans started living in one place. A point was reached where a particular geographical population, or a series of populations, no longer had an unlimited opportunity to survive by past methods. Presumably this could have been a reflection of an environmental change. When this happens it becomes more economic to plant your crops in a restricted area so that instead of having to go wandering for mile after mile, you can gather your crops in one place. Instead of searching for animals you breed them."

Again, and now on this highest of all levels, food is back of it all. I do not quite ask Carr's question: Who first had the wit to know a seed once planted would grow? "So that until agriculture," I asked instead, "all men and near-men going before found their subsistence by other means, and that is the larger meaning of the term hunter–gatherer?"

"Yes. Wild plants and meat, from natural sources on a seasonal basis. You wander over the open country in search of food, you find it where it's growing." Although there are some few still living in such fashion (and irrespective of their instinctive talents), there are no men living today equivalent to prehistoric man for the reason that all of us are of the single species, *Homo sapiens sapiens:* we possess language. Living outside the boundaries of civilization does not class a man as "prehistoric," a mistaken assumption made by Darwin when he first encountered Patagonians. The more accurate term for such people is hunter–gatherers, living by opportunity from the food provided by the land, and still without knowledge or need for agriculture or domestic animals. A most extraordinary example are the Tasadays, the tiny community which was discovered only in 1971, living in total isolation in the Philippine jungles as they have for more than 400 years. The misconceptions about such people are widespread and surely account in large part for the pernicious roots of racism. If you were to introduce an infant from a Tasaday community into a modern Western home, Leakey says, and allowing for the variance in intelligence from one individual to the next, that infant's potential for learning would be generally the equivalent of one born within that home.

"What does that tell us about the relative state of civilization?" I ask.

"It tells me," Leakey says, "that civilization—and/or technology—is purely the product of environment. We do different things because it's more suitable to do different things. You are born into an environment and you learn to cope with that environment." We do not discuss "feral" children—the oft-raised and controversial examples of children abandoned or lost and "adopted" by wild animals, a recurring myth of civilized people. Cut away by happenstance from the world of accumulated knowledge, the child in crisis adapts to the ways of the animal pack, moving on all fours, grunting and howling, bereft of language: the erasure of humanness complete. Recently a New York sociologist suggested a comparable but reverse transformation among city derelicts, the bag ladies and Bowery tramps desensitized to so complete a degree their sole concerns become warmth and the scavenging of garbage. "It doesn't mean one is more primitive than the other," Leakey goes on. "Just that one is adapted differently from the other."

"Then civilization is wholly acquired?"

"Yes. Certainly it's acquired. Technology is acquired. Our behavior in these respects is totally learned. This is an important point. Because your reaction to *stress* in civilization—which is violence—is an acquired reaction, and not innate. If the gene is programmed to be nasty, as some sociobiologists seem to think, then if you put it in *any* environment it's still going to be nasty." He has set himself strongly against the Lorenzian view of human nature, that aggression leading to violence is innate, this having come about largely from early man changing his diet from plants to include animals; killing animals for food eventually leading to his killing his own kind. Instead, Leakey argues that evolution occurring so swiftly had to derive from the social primate's more natural tendency to cooperate; to share. But if this is so, he asks himself, in his book, "Why then is recent human history characterized by conflict rather than compassion? We suggest that the answer . . . lies in the change in a way of life from hunting and gathering to *farming* [italics mine], a change which began about ten thousand years ago and which involved a dramatic alteration in the relationship people had both with the world around them and between each other." For Leakey, the hunter–gatherer is a part of the natural order which the farmer tends to distort. Moreover, farming leads inevitably to the acquisition of possessions, and possessions lead in their accumulation to cities, states, and nations. And since having possessions leads, too, to the need to protect them, "This is the key to human conflict, and it is greatly exaggerated in the highly materialistic world in which we now live." It is some surprise to me that he takes it all somewhat further than either Lamprey or Dasmann.

"And if violence is learned," he says now, "you can unlearn it, or learn something else. Or change the environment. This seems to me very important in terms of the future of the species."

The sun is shining brilliantly on the terrace outside. No place is a good one for feeling bad; Nairobi, because it is so beautiful, is the worst. All the same, my cold gives new, small signs of compromise. "But even if aggression should not be a factor in our behavior," I said, "how do you account for our seeming failure to deal with the amount of food available against the great numbers who need it? If as a species we are demonstrably superior in our ability to adapt by learning, to condition ourselves to virtually any environment, why wouldn't we be clever enough to learn how to conform to the most critical constraints as other animals have managed to do?"

"Because we've always had the ability to move on," Leakey says. "It's only in relatively recent times that our populations have reached the stage where we can't migrate. This was fairly common until there came to be national boundaries. It's only when you stop people moving, as you have to now, that you start to see there is a problem. That's what we're going through now, and that's why we're concerned. Because the *dynamic* has been upset."

I begin again: "But the human as a species—"

"The problem is that very few people think of themselves as part of a species. *That* is the problem. If we could get away from thinking in terms of families, or villages, or towns, or cities, or nations, or ideological groupings and begin to think of ourselves as a species, if we could recognize that the species is threatened . . ."

Carr's paradox (which he insisted was no paradox) intrudes. Intelligence, the victim of its own cleverness; intelligence, its only salvation. "Doesn't this take us back to the brain?" I ask. "You use that loaded word 'think' in there. You say we don't think of ourselves as a species. Other animals don't have to think about such things. They have an adaptive capacity, so that their numbers are set by natural selection to the circumstances that prevail. But you say now it is a thinking process for us . . ."

"It *has* to be a thinking process for us," Leakey says. "We've got ourselves into this problem because of a *thinking process.* By thinking, we've developed technology; we've developed racial boundaries; ideological boundaries. We've done all sorts of things that aren't natural as a consequence of the thinking process. My argument is that we do not continue to do this. A lot of the boundaries we set up were important in history to get us this far. Now it is essential to remove them if we are going to get any further."

"Say that we did—did begin to think the way you suggest. How do you think this would change things?"

"I don't know." There are two waiters left standing at the rear of the dining room; they are not glaring at us but they clearly do not look happy. "But I think the sense of environmental crisis is at least part of the answer. People are beginning to realize that everybody requires a decent standard of living. And most people are sufficiently democratic to accept that that's fair. Now then, you must define what a decent standard of living is. Is it the sort of life followed by most Kenyans? A simple home, no electricity, no power source other than firewood, charcoal, no motor vehicles or public transport, no running water? Or does it consist of five television sets, three cars, three houses, a yacht, and a vacation somewhere around the world whenever you want it? Those are, if you like, the two extremes. It is quite clear that greater numbers would survive at the standard of living typical of the Third World—the resources are there at the moment. But if everyone tried to aspire to the Western end of the scale, it is quite clear, too, that there are too many people. You couldn't maintain them. Both sides have got to shift, to meet in the middle. Now how you bring about a reduction in the standard of living is another question. Presumably this will emerge from a realization that the resources now enjoyed in the West are increasingly obtainable only *outside* of the West. Utilization of these resources will have to proceed on an equitable basis. If you take them by force the consequences will be nuclear war. So if you're not going to take them by force, you have to take them peacefully, and that means we will reach an understanding which will result in a shift—a reduction of the consumption on your part. And this in itself will lead to an improvement."

11

THE RED CAR

Trout Lake, Washington

Winding his watch was a must. It held him to his schedule. Once a few days ago he had forgotten to do this, and he was really upset with himself until he realized his sense of time had become quite as remarkable as his ability to hear what he was listening for while asleep. Without looking at the watch, he knew almost to the minute when the sun came up, when it rose over the trees, bringing warmth into the ravine, and when it went down. So he had simply waited for sunset and turned the hands to 7:30 P.M. He was certain he couldn't have been more than two or three minutes off.

That done, he combed his hair, and then he scrubbed his face and hands with the wet T-shirt. Today, Thursday, the twelfth day, would be different. He had himself under control now. He was absolutely clear in his mind, he knew exactly what he must expect of himself. There was nothing he could do about it yet—although his face was still swollen, his suspicion of yesterday that it was starting to go down had been correct, it *was* going down—but there would be plenty he could do once that was a fact, once he had back full peripheral vision and could see back to his pinned foot so that he could start work on it again.

Meanwhile, the thing to do was to forget the unhappy events of yesterday and look at the positive side of his situation. He had water and he had shelter. The gasoline spilled on the ground beneath him must have evaporated; the fumes were gone. So he was safe. He had no hunger, not the slightest bit. He had the resources that came built into the wagon: wiring, rods, cloth, metal fixtures, rubber, plastic—and he had the tire iron. Most of the time he was warm enough, and some of the time he was dry. When he lay absolutely still, although admittedly this couldn't be for very long, he was without pain. He had no fever, and if there was gangrene in his foot—probably was—it hadn't begun to spread up his leg. He had his strength and with the obvious exception of his foot, he had his health. And now the swelling in his face was going down.

He wouldn't let it matter so much about the trucks. Already now they had begun to rumble over the gravel road, slowing to gear down—slowing near the exact spot they should have been able to see Vihtelic in his wreckage home. Should have but hadn't. He knew now there were three of them, that they began up the road toward Hood, rolling down and around the turn, shifting down before the bridge, rumbling over it and heading on a ways toward Randle, and then turning back again. He recognized each of them separately, estimated five round trips apiece, and had concluded early yesterday why they were there: they were working to improve the road connecting Rainier to Hood.

Forced by the weight of their load to move slowly over terrain that would soon become routine to their drivers, even at one point gearing down to a crawl, they virtually hovered in the air above John Vihtelic inside his wrecked car which was clearly marked by his bright rugby shirt and, when the sun finally came out, by the piercing light from his racket-mirror playing along the trees about them—and, for one brief instant, on the cabs of their trucks. *Why don't they see me?* Vihtelic had wondered.

But for most of yesterday, holding to his routine to fill the intervals, he hadn't let it get to him that they had missed him so far, being buoyed instead by his obvious good fortune in the circumstances which brought them there—until late in the afternoon when the red car came. He heard it as he had heard the others, the quiet engine humming above the stream well before it would arrive at the break in the trees above the bridge. Vihtelic got his racket ready, flashing in on the opening to get an exact fix, his pulse throbbing as it always did with another new chance. He saw the redness of it through the trees as it moved toward the opening. It seemed to be slowing. Why? he wondered. Slowly it nosed into the opening on the bridge above the trees. It stopped. It was in full view of Vihtelic. *He couldn't believe it!* Was he hallucinating again? Flashing with the racket, screaming at the top of his lungs, he nevertheless sought some distance from himself to be sure he was awake and fully conscious, and not dreaming again—that this red miracle of his salvation was truly there, on the bridge above him. Stopped. Vihtelic screamed and banged on the car chassis with his tire iron. He dropped the iron and flashed his mirror, still screaming, and then he grabbed the iron and beat on the chassis.

When the red car moved on, he thought for a long time afterward it had been there for minutes, long minutes—for a long time. But really it had been there only a few seconds, he realized that now. But it was certainly no dream. It had stopped and it had gone on, leaving Vihtelic in a pretty bad way. Vihtelic had scuttled back into his steel shell. He remembered that the ravine had looked unusually beautiful to him, the green dark trees reaching high above him, the warm breeze brushing past him. The birds were bathing and drinking from the stream. One fluttered quite close, seeming not to notice him. Vihtelic's beard had grown out; his face was dark within the darkness of the car. It was unlikely the bird was aware he was there. He forced himself to admit the harsh truth that came with the red car.

The truth was that he couldn't be seen from the road above him. If the driver of the red car hadn't seen him, no one could see him. Neither the drivers of trucks nor an army of searchers. For the first time Vihtelic thought that he really might die, and if he did, this might not be such a bad place to die. It was green and shaded, with the white stream flowing down the heart of it. He had written his fourth letter to Mary, and then he had thought to add on something about how he wanted his possessions divided.

But suddenly and quite sharply he had realized he was not thinking right about his situation, that he was about to lose his sense of organization. He forced the paper away from himself with revulsion—threw it out of the car—and resumed his chores. The point he had almost missed yesterday was the point he would live with now, until he was out of here. *Nobody along the road could see him, which meant nobody was coming down here to get him out.* He studied his face in the mirror. To hell with the damn trucks. The swelling about his eyes *had* begun to recede.

The Pleistocene

In a cave called Shanidar situated in the Zagros Mountains of Iraq, 250 miles north of Baghdad—a shelter now used by the Kurds to protect their goats from the weather—researchers discovered the remains of a man about forty years of age, five feet three inches tall, who was buried on a June day 60,000 years ago within a bower of flowers—grape hyacinth, aster, hollyhock, bachelor's buttons, and St. Barnaby's thistle. Photographs of the fossil skull of this man show it rising from the stone rubble in which it was found, like Rodin's head of Balzac, the massive brow brooding over the times its eyes had seen. The remains of the flowers were pollen grains and they had been arranged so discretely as to suggest the formal observance of a rite. By 60,000 years ago, then, if not earlier, it is assumed there were human feelings: of loss and grief, of prescience, and the awareness of death as an end to living things.

From the fading distance of *Ramapithecus* toward the closer past of the man of Shanidar, apish men slowly worked their way beyond apish ways. As the brain grew larger, the hand changed. The power grip by which the fingers enfold an object to grasp it—a characteristic men share with apes and monkeys—was augmented by the precision grip, the longer and stronger thumb of man countering the fingers to permit a finer manipulation of smaller objects, a dexterity. (An ape hand can turn the handle of a screwdriver but it cannot first fix the screw in place.) Crude tools became complex. From the stone wedge of the earliest pre-men, there were by the time of the man of Shanidar sixty tools which were used for many purposes. As tools evolved cultures began: communities formed; men hunted together and shared their food.

By 40,000 years ago modern man had emerged, and so suddenly that by any reckoning of geological time, there is no wholly satisfactory explanation for the fact. Though the transformation occurred over a period ten times longer than the recorded history of civilization, so much happened so relatively quickly that the event is regarded now as an evolutionary explosion—"quantum evolution," as some describe it. From 40,000 years ago through the next 30,000 years men built shelters for themselves, boats, fishing implements, weapons; fashioned art through paintings and sculpture; made costumes; and they moved about the world into even the most inhospitable of places, like the Arctic.

Possibly the most important of many important reasons for this was by now the new man's capacity for language, not just the ability to express thought, which is not unique in the animal kingdom, but the ability to communicate it directly, which is. Through the ability to express their thoughts—to exchange them among themselves—men were able to tell their children how to do things other men had done before them, and the more formal accumulation of knowledge began.

By 40,000 years ago, pre-men had become human beings. They looked then as they do now. They lost the heavy hair that had covered them (to some advantage: the human skin contains five million openings to the body's sweat glands, facilitating through quick evaporation a sudden action response even in the most intense heat—and to some disadvantage: men accordingly need more water to keep going). And though there would come to be superficial differences in the color of their skin (the protective pigmentation varying according to relative location to the intensity of the sun) and in physiological structure, such differences caused by adaptation through separation of groups in different parts of the world, they were all of the same species, or, more precisely, of the same subspecies, *Homo sapiens sapiens.*

By 10,000 years ago, when the great ice sheets had melted and the seawaters had risen, some humans developed methods of control over various plants and some other animals, by growing and raising them to provide food in stationary settings. They did this through the establishment of agriculture—the domestication of plants and of selected animals, both events occurring at roughly the same time. The human population then is believed to have been less than ten million. While many of that number had settled about the eastern shores of the Mediterranean, where it is believed that agriculture first began, many others continued to live as they had in the past—hunter–gatherers foraging for food wherever it was to be found. Of these latter those who today still survive by such methods number less than 30,000 and are located in the most remote corners of the earth. The rest of the human population at present numbers four billion, living in ways of varying accommodation to the consequences of the emergence of agriculture 10,000 years ago.

Canberra, Australia

Corn comes from wild grass. People living in the Tehuacan Valley in Mexico domesticated it five or six thousand years ago into the vegetable known as maize. The Pilgrims of New England were shown how to plant it by an Indian named Squanto, and the early settlers, moving on across the country, spread its seed wherever they went.

Corn is eaten as is, or it is ground into mush or grits or leached into hominy, or it is used as a meal base for tortillas and tamales. For domestic animals alike—cows, pigs, and chickens—it is a primary feed grain in both the United States and Europe, where, at its most efficient rate of exchange, two pounds of it are needed to produce a pound of chicken. Cooking oil comes from corn, and so does starch, which is used in everything from paint remover to high explosives; and so does sugar, important currently as an additive in the form of alcohol to gasoline (gasohol), a stopgap measure to help extend the United States' fuel supply.

A man named Donald F. Jones is largely responsible for the fecundity of modern corn. In 1917, from his experiments in breeding selections, Jones developed a hybrid strain which was shorter in the stalk, which could be placed in rows closer to one another, and which responded more efficiently to fertilizer.

The immediate consequence of this was a dramatic increase in production. By 1929, the national yield had gained by 25 percent, and soon thereafter Jones's hybrid corn had become America's boom crop. Whereas farmers previously had grown their own seed, the new methods of artificial selection now required the skilled manipulations of technologists. (But at the time, for those who may have paused to reflect on the loss of personal control, the remarkable proliferation of the plant seemed more than adequate compensation.)

As it became less expensive to produce, Jones's hybrid corn became cheaper and more available, and soon a mass market arose to accommodate the buying, selling, and distribution of it. In turn, the mass market helped to shape the further development of hybrid corn. Mass markets favor products that are uniform in size, shape, and availability; for produce to become uniform, there must be uniformity at its heart—that is, in the genetics of the produce. Through techniques of selection devised by Jones and others coming after him, modern corn plants became genetically as similar as identical twins. The advantages of hybrid corn (and other hybrids such as sorghum, sugar beets, carrots, and onions) were such as to establish the United States as first in food production around the world. By 1973 food exports returned twenty-one billion dollars to the U. S., more than that of computers or jet aircraft, or any other kind of high technology.

For the human species among all animals, the mass production of food over the last hundred years is an achievement of great brilliance, seeming on the face of it to have countered the deadly prophecies of Thomas Malthus. As human population has steadily climbed, food production has stayed ahead—even through a time when human numbers threaten the planet: two billion people by 1930, three billion by 1960, four billion by 1977, and at the annual growth rate of 2 percent, six billion expected by the end of the century.

But through the intervention of modern technology—new breeding techniques, heavy machinery, the infusion of electrical energy, and improved fertilizer—food production has grown even faster than the numbers of people, at the rate of 2.5 percent a year. Food, the spur to evolved life on the planet, is a problem that would seem to have been uniquely solved by man in general, and specifically by such men as Donald F. Jones and his new strains of hybrid corn.

In the spring of 1970, however, something happened to raise doubts about such assumptions, at least insofar as the success of hybrid corn was concerned. The dominant strain then, having been bred from Jones's hybrid, was something called T (for Texas) cytoplasm. Cytoplasm is the substance which surrounds the nucleus of the cell in all living things. Having been bred to the specifications of the mass corn market, T cytoplasm corn was spread about the country in billions of rows covering hundreds of thousands of acres, the cytoplasm in the cells of each plant a precise replica of that in the cells of its neighbor.

In Florida, where the growing season for corn begins, suddenly and without warning, corn began to die. As weather warmed to the north and the growing season moved up the country, the death rate spread. By the end of the growing season, a fifth of the corn crop of the largest corn-producing country in the world was a total loss.

Only a few people at the time had a clear understanding of what had gone wrong. One of these was Sir Otto Frankel, chief of the Division of Plant Industry, Commonwealth Scientific and Industrial Research Organization (CSIRO), in Canberra, Australia, who, by prearrangement, is supposed to be among the handful behind the terminal gate, awaiting the debarkation of Flight 273 but not nearly so eagerly as I, inside Flight 273, waiting in the aisle to get out. I have no idea what Sir Otto looks like and am too bleary to care much. But he doesn't know what I look like either. At the gate I hang back from the others and work at expressing apprehension—evidently the right signal, for a tiny man with white frizzled hair detaches himself from the others and moves swiftly toward me. The first impression of him is otherwise that of a fast-moving blur.

"Sir Otto?"

"I am *Otto,* which is the way we do it in Australia and America. Let's hear no more of *that,*" he says, wresting one of my bags from me and striking out across the terminal toward the reservations desk. I must confirm my return flight, he says over his shoulder; it is always a good idea to do it first thing upon arrival. The clerk says the flight out Saturday, the day after tomorrow, is uncertain because at this time of the year—about noon when I am to depart—there could be the problem of fog. "No," Otto says firmly, ignoring the man. "There is a low center coming tomorrow and it will rain. That means it will be clear Saturday. You will have no trouble."

On second impression: he wears fashionable speed-goggle glasses, which give considerable magnification to his eyes; he is nattily got up in tweed jacket and ascot; and his accent is upper-class British scrambled by the vestiges of a Teutonic past. His *r*'s turn to *w*'s, making bumps in British idiom. ("He is weally a *vewy* odd chap, you know.") As my wits slowly coalesce, ordered to attention by his unrelenting exuberance, I see finally that he is somewhat older than he first appears.

Twenty percent of the world's uranium is located in the central wasteland of Australia, Otto says as he zips me away from the terminal in his small car, and there is currently a hot controversy in this country, considering the possible consequences, as to whether it should be mined. Incidentally, this disputed area is within the vicinity of a past habitat of the Aborigines, who invest in all of nature—even rocks—their spiritual ancestry. They have filed legal claim against the miners for posing a threat to their sacred sites which date to times beyond written record. Well, anyway, he wants Lady Margaret to go to a lecture about it at the Australian Academy of Science—by Lord Bulloch; his topic will be "The Faustian Bargain"—although he himself won't be able to make it since he has to clear his desk for his vacation. He is going to the mountains for a week of skiing.

At seventy-eight, Sir Otto Frankel tells other nations, and the United Nations, what they ought to do to grow more food, although he himself wouldn't go that far; he is in fact at pains, even to the point of pedantry, to put things precisely as they should be, which makes him not an easy man to talk with. ("Now, hold on! I can't let you overstate these things even though they describe the position I hold!")

Frankel's specialty, genetic conservation, has made him important to an understanding of, among other genetic enigmas, the odd and alarming behavior of the American corn crop of 1970. What went wrong then was related essentially to what went wrong in the forties and fifties when an antibiotic would suddenly stop working against the disease it was intended to cure, or a pesticide would stop killing pests. The understanding of the problem lay within the nature of evolution by natural selection—on the cellular level.

In food plants the cytoplasm of the cell is, in addition to the genes, an important aspect of diversity. Although its composition varies between species, the cytoplasm in T cytoplasm corn was—to the extent that it determined susceptibility to blight—the same wherever the strain was sown, which in 1970 was across most of the country. Present also, wherever *any* strain of corn is sown, is a fungus blight called *H. maydis*—under the microscope it looks like tiny worms—but it is seldom much of a problem since its more virulent strains are usually checked by genetic variables. Except, again, in 1970. As T cytoplasm was being refined, *H. maydis* was mutating to live. The uniformity of the cytoplasm in the 1970 crop set the crop up for a killing, and when one of the mutants of *H. maydis* met no resistance, disaster ensued. As Hugh Lamprey had warned, arguing for diversity, "some lucky predator, some pest" was offered the feast of its life.

The corn blight of 1970 was the direct result of genetic vulnerability through the uniformity of T cytoplasm corn, the kickback factor in food production—the elaborately delicate tracing of such inexorably logical behavior being the special expertise of my new friend, Otto Frankel.

He left me at the hotel with the promise he would be back to collect me at six for dinner at his house. After some seventeen hours in the air from Los Angeles and the circadian confusions resulting therefrom, I had fallen on the bed like a rock, coming to only a few minutes before he arrived. In the car, as though he had been talking since he left me, he goes on skipping blithely from one subject to the next—Canberra was laid out by an American, Margaret and he took a four-day hike through New Zealand, there is a tree outside his office so beautiful he can hardly write when he becomes aware of it—and on through dinner, through brandy later, and on along the drive next day to his office at CSIRO. His eyes are interesting. The speed-goggles enlarge them, and they droop slightly when at last he sets himself to listen—to wait, rather, until he is ready to get back into it. When his respondent draws toward a point, the lids slip lower still, snapping open suddenly, like sprung shades, a warning he is about to set the record straight. There is an assumption enough food exists to feed people indefinitely? His voice trembles with exasperation and his eyes fly open, wide as a tarsier's.

To relax the initiative is to invite its capture by Otto Frankel, who has his own curiosities to explore. What was I up to anyway? Where had I come from and where was I going next? What did I want to find out—not only from him but from the others? What had they said? What did I *expect* them to say? Why was I starting so far back with it all? Didn't I know I would be getting myself into the complexities of chemical compositions? Why deal with the clumsy, uncertain business of where animals come from? How could people without scientific backgrounds be expected to follow such things?

When I venture that I am precisely such a person, the sort I would hope they and he might inform, his eyelids begin to droop. He waits for more. But what more? The impression swiftly comes that his reassuring cordiality ends abruptly at the edge of his scientific interests. The others told me agriculture started 10,000 years ago, I say.

"We don't know why." This is said with some defiance. Watching me at half mast, he seems to expect a challenge.

"We don't?"

The eyes open: "Because early man didn't need it! The idea of people living then as hunter-gatherers in a state of extreme poverty, working terribly hard for their living, is completely disproven by the life of the few peoples who live as hunter–gatherers today. Ancient man *didn't* need agriculture! The Aborigines of Australia have *never* invented it. There was no population pressure back then. Jack Harlan"—he is a crop evolutionist at the University of Illinois—"writes about all this very well, and he describes those times as man's golden age. Man worked very little, his food was easily gathered. At that time he probably had his most trouble-free existence."

He waits again, signaling something. Another gambit? I wait. "Harlan believes that agriculture started not in the great river basins where it flowered eight thousand years ago or less—in Mesopotamia or India —but in areas adjacent to them where life was tougher, and where population pressure forced men to cultivate." Most of these places, he says, were located in mountainous areas near the tropics, on steep slopes of varying climates. Hard scrabble. Whatever managed to grow there has had from the first to fight to stay on. The earliest food plants are still there—or were, until recently. A pioneering Russian geneticist named Vavilov in the twenties began systematically to track them down, collecting seeds for storage from each of them. They are known now, in his honor, as Vavilov food centers. (Vavilov was himself taken from a grain field in Poland one day by Soviet authorities, who disagreed with his genetic theories, and never heard from again.) In many instances he found farmers tending these primitive plants as they must have for thousands of years in the past. Because of their heterogeneity and the simple fact of their endurance through millennia, they were recognized by Vavilov as a vital source, a gene reserve for the sustenance and enrichment of modern plants, and as resource for protection against plant disease. These ancient plants—living fossils in their own way—are called "land races," a term I had never heard, and my confession of unfamiliarity merely causes Otto's eyelids to slip down again.

"Would you explain for me exactly what a land race is?"

"I'll give you a Xerox of a chapter I've written on all that. It's more economical and more thought-out," he says, shuffling his papers to put before me a paragraph which is, as promised, admirably compressed. But it is not so vivid as one I later found by the American biologist Garrison Wilkes:

> In many instances the wild plant from which the domesticated crop was derived is no longer extant. In these cases, the varieties closest to the ancestral form are the primitive land races still being grown in those regions of the world where the plant was originally domesticated or in regions where the crops had originally acclimatized a long time ago. These regions are the genetic resource areas, or living bank accounts, to which the plant breeder turns for additional germplasm in the plant improvement process. . . . These reservoirs are our irreplaceable genetic heritage which has assured freedom from want and hunger for every civilization that has ever existed.

"Land races," Otto says, his eyes widening, "evolved wherever the earliest domesticates migrated. This is terribly important to realize! Because the diversity of land races depends on—was *created* by—the diversity of environments to which a crop migrated. A few crops stayed out. Teff is one of them, a purely Ethiopian crop. It has a *degree* of genetic diversity because Ethiopia is a mountainous country and therefore there are altitudinal ecological transects, soil differences, and so on. So it diversified there. But had it spread about the world it would be very much more diverse."

This was certainly the case in the movement of food plants across the United States, as Wilkes points out. Immigrants brought plants and sowed them wherever they settled: barleys from Germany and England; wheat from Calcutta, rice from China, different strains from widely varying international environments which mixed and spread over near-virgin soil, creating a broad genetic base for the plant breeding of the future—the literal reality within the myth of Johnny Appleseed.

"In the temperate zone," Otto goes on, "a majority of cereals and pulses—grain, legumes, peas, lentils, and so on—originated in the fertile crescent in the Near East and came very early to Ethiopia. In Ethiopia, barley developed more genetic diversity than anywhere else in the world. Ethiopia is really the *fountainhead* for genetic diversity!" He stares at me, wide-eyed, without blinking. "For example, there is a resistance gene to a virus—barley yellow dwarf virus—and the only resistance gene to this virus in barley comes from *there.* " He stops and stares again. "A good example of the value of land races! If the Ethiopian land races of barley had been displaced *before* barley yellow dwarf spread around the world—it was not so very long ago that it happened, in this century—if they had been destroyed earlier, we would never have gotten any resistance."

With Mendel's discovery of the heredity principles in garden peas, plant breeders like Donald Jones were afforded a technique for the swifter acceleration in the improvement of food plants. Traveling around the world to Vavilov's food centers, and to the places where the earliest foods had migrated and adapted, they built a living file of genetic supplements to increase through crossbreeding the yield of plant foods.

In time a remarkable attitude developed among these scientists. They came to believe it might be possible to grow enough food for all the people in the world. By modifying and crossing strains, new food resources might be developed outside the great farmlands of America, Canada, and Argentina, even in those places where much of the first food came from—lands that now were barren. By the late fifties there were radically new strains of wheat and rice that could be made to grow almost anywhere, so long as there was sufficient water, massive quantities of nitrogenous fertilizer, and pesticides.

By the early sixties, among these food scientists, there was considerable confidence that worldwide starvation would soon become a human tragedy of the past. One of their number, a geneticist named Norman Borlaug, received the Nobel Peace Prize for his part in devising what by now was called the Green Revolution. So dramatically effective were the new artificially selected seeds that farmers all over the world began to favor them to the exclusion of all else, including the more natural seeds they had always sown in the past—which necessarily included an unknown number and variety of land races.

And so it was against these optimistic international developments that the failure of the corn crop of 1970 caused such shock, the extent of it moving up in levels of concern—first, among the American farmers who simply lost money; next, among breeders who found themselves outflanked by *H. maydis,* an ordinary fungus of unimpressive credentials; and lastly, but most seriously, among geneticists like Otto Frankel who saw the danger as different not in degree but in kind. Potentially, genetic uniformity in crop seed threatened an evolutionary disaster.

Wilkes again:

> The southern corn blight was a dramatic demonstration of the inherent danger of genetic uniformity. . . . Such a crop failure would have been disastrous in countries like Guatemala or Kenya, where corn is half the daily caloric intake.
>
> Yet such disasters have occurred. In the eighteenth century, a new food plant from the Andes of South America, the potato, was introduced into Ireland. The genetic diversity of the introduction was small; but, isolated from some of its diseases, the potato yielded well and the Irish population increased. In the 1830s, with the population having increased threefold to eight million, a previously unknown disease caused by the fungus *Phytophthora infestans* appeared in Ireland. Within two years, two million Irish emigrated, two million died and four million remained, many in abject poverty. The Irish had inadvertently narrowed the genetic base of their crop. . . .

But by the mid-seventies there was even more trouble. Sharp increases in the price of oil began to drive up the cost of fertilizer, and fertilizer is essential to make the high-yield grains of the Green Revolution grow. Quite suddenly, for impoverished countries now dependent upon the new techniques, a vital ingredient to the package of agricultural innovations out of the West had become severely limited and therefore prohibitively expensive.

Nevertheless, on the surface of the times, the astonishing increases in the rates of food production in just a few countries, especially in the United States, raised hope among even the most cautious of observers. Between 1973 and 1975, for example, despite the turmoil created by OPEC, the United States accounted for 65 percent of the world's total grain exports while requiring only 5 percent of its labor force for the production of *all* its agriculture. There was such an abundance of it, in fact, that only 16 percent of the disposable income of the average American was spent to provide it. Even during recession years, food had become a growth commodity, adding richly to the nation's foreign exchange.

Still, all of this was on the surface. In an address to the Markle Foundation (and not circulated much beyond it), Phillip Handler of the National Academy of Sciences warned that the production of world food at the annual growth rate of 2.5 percent, as against a population growth of 2 percent, simply did not provide enough food to go around. Above the equator people were well fed for the most part; below it, many were not. Indeed, by the end of the seventies, one in eight of the world's people would suffer malnutrition to the extent, among children, of possible brain damage. The production and distribution of food were governed by interests beyond the simple satisfaction of need. Food gave its owners a political edge. Moreover—the Malthusian principle holding—the true need was for a *3* percent growth of food against present population increases, and all indications were, irrespective of the continuing exponential human increase, the rate of food production had by now reached its peak. The land—the arable part of it—was giving out. (In America, in fact, where farmland was most plentiful, the Council on Environmental Quality reported that the amount available was shrinking through conversion to other uses at a rate through the seventies of three million acres a year—a projection which, if sustained through the year 2000, would fall to zero.)

Losing what may still be there is mostly what worries Otto Frankel. Through the rest of the day and into the evening, he talks of the complex nature of genetic conservation.

"Look, I want to come clean about something," he says at one point, fixing me with his stare. "I said that land races had not been very extensively used. It's very much handier for a breeder who wants a reasonable level of productivity to intercross advanced cultivars—modern food plants—than to cross them with a very low-productive land race." He looks hard at me to see if I have understood. "Land race agriculture in the modern world, with all the investments we have to make, would be just inconceivable! Besides, the world would be starving. Without plant breeding and fertilizer the whole *world* would be starving." Land races, in short, could not produce the massive supplies of food now needed. "You might say there was little fertilizer prior to 1850, and there were no breeders' varieties at all. This is perfectly true"—the eyes widen—"but look at the population size then!"

Because it is the most productive way, the crossbreeding of modern plants—one with the other rather than with their ancient ancestors—is the modern way, and one more Faustian choice. Such as there are left, land races are living insurance against problems unmet. They are less important as food than they are as food's protectors. But they are insufficiently identified, poorly cataloged—and, through their recent displacement by modern seed, which is needed to sustain the existence of more and more hungry people, they are rapidly disappearing. "It is like taking stones from the foundation to repair the roof," Wilkes says.

What is required to save them?

"The only possibility, and that is being done, but inadequately," Otto says, "is to get representative samples in the field. You go to the areas where land races still exist and over a period of two or three years you sample the genetic diversity. Then the results have to be safeguarded under adequate storage conditions." He makes it sound easier than it turns out to be. Land races are not so readily identified even by scientists, let alone by those who are innocently destroying them. It is thus a race against both time and ignorance. Collectors visit public places in the Vavilov food centers, frantically buying seed in street markets to preserve strains from extinction. One of them, Dr. Erna Bennett of the Food and Agriculture Organization, recently drove her own car from the Aegean Sea to the Himalayas, storing in the back seeds which hadn't changed over the past 2,000 years. She arrived home with half a ton of them.

"Most seeds that are properly harvested and stored in a good condition have a fairly long storage life—decades rather than years," Otto says. "So it is possible to store very large numbers of seed samples in a relatively small place. Theoretically one has to make germination tests, and when germination drops, then one has to grow the sample and rejuvenate it. And also one has to have available a stock of seed for distribution so that plant breeders can get it, as can institutions which want to test a range of samples for anything at all —disease resistance, quality characteristics, chemical content, nutritional quality, and so on."

"How many would you need of a given seed?"

"It depends on the size. One kilo, perhaps a kilo and a half. Various procedures have been worked out but they are hardly in existence, because there are very few places which have the temperature range and such a regime of maintenance. Very few in the world—Fort Collins, Colorado, for example, where the American storage center is, has some defects. It's the oldest and by far the largest in the world but it suffers from a shortage of funds."

"Who supports it?"

"The U. S. Department of Agriculture."

For an instant I recall that Fort Collins is where I left Evans worrying over the rights of wildlife versus man's mineral rights.

Sir Otto Frankel is a reluctant but, once pressed, eloquent sentimentalist, appealing to emotion when reason is too complex to get across quickly. At the Stockholm conference on the Human Environment he stirred the assembly by devoting his allotted three minutes to the consequence in human terms of continued indifference to problems of environmental crisis: "You have a responsibility not just to yourselves," he declared, "but to your grandchildren." But however deeply felt—and it is not at all inconceivable that the tree outside his office window could move him from action to reflective silence—his sentiments are securely guarded. His manner is that of a hard-edge, no-nonsense, practical-minded scientist who is quick to distinguish between those problems which are manageable and those which are not:

The hazard of plutonium pollution is manageable even though its harmful radiation has a life of 35,000 years; places will be found—such as the center of the Australian desert—to dispose of it.

The disappearance of animals as a food source is manageable; the disappearance of plant food is not. In one form or another, animals are parasites of plants, and there is *nothing* that can replace plants.

The destruction of a species is manageable; the destruction of habitats is not—because the possibility of the restoration of life in any form is then lost.

"My justification of conservation is that it supports the preservation of a residuum of evolutionary potential," he says without blinking, "because life cannot continue without evolution."

Tomorrow at noon, there being no delay from fog as I am sure will be the case since it isn't hard to trust the careful judgment of Otto Frankel, I will head west on a longer trip than the interminable flight here. There is much to ask this man who is so willing to get things straight, and there are only a few minutes left. He seems by now somewhat more relaxed, so I decide to junk caution and draw from him what I can.

"Someone told me all the trees in China were cut down three thousand years ago . . ."

"Yes!" The eyes pop open. "China now is the biggest tree-planting country."

"Why were the trees cut down?"

"To build houses, ships, bridges—and the people then didn't replant."

"But they are replanting now?"

"Because this is a new world, and they can do it. If the Indians wanted to do this, they would find it exceedingly difficult. The Chinese authorities just say, 'Do it.' "

"Is there a potential food shortage there, considering its immense population?"

"No. The agriculture is highly effective, and they are extremely practical in using research results. The Chinese are strong on propaganda encouraging birth control," he goes on. "There is very, very strong moral pressure from within the community."

The *politics* of a country does not seem to interest him much. He has advised the Indians on their food problems, but in contrast to the Chinese, their methods fail to address the problems they face. The United Nations is of little help. At least, I say, it tries to bring international scientists together. "No, it does not," he replies sharply, glaring at me. "It brings politicians together and they accomplish *nothing.* They merely blow off hot air about the way they want to appear to others!"

I gather it is the *spirit* of conservation that he worries over. In one of his monographs, he writes: "The possibility of a virtual end to the evolution of species of no direct usage to man raises questions of responsibility and ethics . . ."

I ask him about this: "Isn't it clear now that more species than we know are important to our well-being? Why should this be so elusive an idea to grasp? Why the need for an ethic?"

"Because it is a kind of religious attitude that people are not susceptible to. It's not very easy for man to accept a credo unless it's imposed on him by social dictates, which of course include religion."

"But the *dependency* is there. That is a factual situation—hardly a religion."

He says, with some relish, "It *isn't* a factual situation. The implication of my passage was about species that we are not literally dependent on. This is *philosophical.* Do we want a world in which there are no living things except those that we use, or can't get rid of? If you think this out it's a rather horrible world. Have you ever seen northern France and the Belgian countryside? The trees were destroyed in the first war, and not very many were replanted. Treeless now. You have large stretches of land growing sugar beets, wheat, and barley—completely utilitarian.

"I had a short study time in a German university town where every square yard was used by man. The trees were numbered, counted, measured. The fields started within an inch of the roadside—nothing was there that was not used by man. I would find it abhorrent to think that nothing exists that I don't need or can't live without. But I have put this to other people and they seem never to have thought about it—so unreal to them that they needn't bother. But it's a very real possibility that the trees and flowers in a public garden will come to an end. Living in New York City, for instance, I find a ghastly prospect. But if I say *that,* someone will say, lots of people are very happy in New York. Not all the eight million are unhappy. This is obviously justified. They have made an adjustment of the social psyche, which spells doom for wildlife."

Whose views reflect his own in this concern? John Passmore, he says. E. F. Schumacher. The first is not familiar to me though the last, as author of *Small Is Beautiful,* is. But I haven't read his book, I tell Otto. "It is interesting and important," he says. "People say they don't understand how he proposes to accomplish his methods economically, and he says such reservations are rubbish. They *can* be accomplished."

"Do his models include industrial states going back to smaller economies?"

"It is not a matter of going back, it is a different kind of organization he has in mind. How small can an efficient plant be? A tire plant, for instance, can be quite small. New Zealand needs only two. There's a *lot* in that. What has changed things in our time is the power of the multinationals. They of course have an interest in the big rather than the small because the big is more easily managed—from their point of view. Although you *could* have small factories belonging to multinationals. . . ." He allows himself a long sigh. "This is a difficult world, and we have really stacked the cards so heavily against ourselves! Probably we have *always* done that, but now the system has become so complicated. In some ways maybe we'll muddle through. There is a quotation Schumacher has at the end of his book—that no problem is ever solved, but we find some way of dealing with it. Not solving it, but just nibbling at it."

It had not been in my plan but I decide now that before I'm through, if I can, I ought to try to see Schumacher. By coincidence, Schumacher's name has been in the papers here in connection with his efforts to persuade Australians to stop cutting down the eucalyptus trees in the western part of the country. Frankel is careful not to exaggerate. Exactly how important, I ask him, does he believe this component of the world's rain forest is?

"Rain forest is for the most part on poor lateritic soils," he says. "Structurally poor soils, poor in humus. The humus is lost through oxidation and bacterial action. That is replaced by the leaves which fall off the trees, and by the trees that rot and fall to the gound, so there is a continuing cycle. Some nitrogen is replaced by rain, and then there are many nitrogen-assimilating microorganisms associated with the trees, or just freely present within the soil. There is a subtle but very tenuous balance which is easily upset, and the removing of the forest is a very grave interference. The Southeast Asian rain forests are certainly the last reservoir for new species which man may wish to use or domesticate—especially in the chemical and pharmaceutical industries. There may be a range of new species which one may want to use which will have vanished unless they are preserved in forests."

"To what extent are these forests under pressure?"

"Whitmore, the principal writer on such things, says that in large parcels, especially in Malaya, there will be no natural forest left by the end of the eighties."

His view of all this is conditioned by problems existing on the other side of the world from mine, but it is apparent by now they are evidently the same problems everywhere. We sit together quietly, and for once he is not poised for the offensive, not daring me to respond. Through heavy-lidded eyes he peers at me.

"There has never been a catastrophe like this since the Ice Age," he says, "and the Ice Age didn't affect the tropical flora."

"But isn't it possible that some of these problems would come to be realized—in the Indonesian system, for example?"

"Of course, they are seen now, but it's too difficult to do anything about them."

"Business pressures?"

"Exactly."

12

THINKING IT OVER

Trout Lake, Washington

By Friday, the thirteenth day, he had changed his strategy. Not that he had abandoned completely the possibility of flashing the trucks—he still grabbed for his racket mirror when he heard one coming—but his face had gone down now, as he had expected it would, he could see back to his foot pinned behind him, and all Vihtelic's energy had become directed toward a different possibility. He had to get his foot free from the dashboard. Some*how.* That was all there was to it.

But how exactly? The best he could think of was to resume the method he had used twelve days ago, prying and wedging with the tire iron between the dashboard and the tree root. With each separate move, as it always had, the pain flowed upward through his leg and into his chest. But the dead foot was another matter. The cost there had lessened considerably, so that he could work about it longer than he had been able to do before. This alone encouraged Vihtelic to believe the wedging action eventually would pay off.

So he had spent the rest of the day this way, and most of the next, twisting and prying with the iron, interrupting the sequence when he heard another truck coming, or when he needed to bring in more water. That was, of course, his constant need, one which never—even by this late date, and no matter how skillful he had become at it—

would he let himself take for granted. It was too critical to his survival. Too many things could go wrong. The line could snag or, worse, it could break. He had only so many shirts to adapt to the purpose, and his supply of wire obviously was limited. Sometimes it took him as long as five minutes to work the wet shirt back to the car. To relieve the monotony and tedium of it, and to keep his attention focused, Vihtelic had begun running a sort of contest between himself and the natural obstacles impeding the progress of the operation. Vihtelic traveled a course run between logs and rocks, and often over them; and the wet shirt, successfully retrieved, was the trophy. Chanting his verse, he would throw it out, then set himself a smooth and unhurried rhythm in working it back. Each drop was another victory.

On he worked in this way, through all of Friday and most of Saturday, wedging and fishing, wedging and fishing. On Friday there were still the trucks, and on Saturday the tourists returned. So what? If they saw him, fine. He would still flash his racket toward them, but he expected very little to come of it. As far as his emotions were concerned, he supposed he was about as well off as he had ever been. His energies and intentions were directed. At noon on Saturday, feeling his thighs and his stomach and gauging the condition of his foot, he guessed he was good for another two weeks down here. At least. But he wouldn't have to wait anywhere near that long, because he was going to get out of here. He *knew* it, and his resolve kept him in full concentration on the problem before him, wedging and fishing, wedging and fishing—on through late Saturday afternoon. Stubbornly he worked until long after the warm breeze had passed, and into the early evening as the sun edged down toward dusk, approaching the end of this second week he had been trapped here, in the wreckage of his car, here at the bottom of this hole. He was somewhat weaker now than he had been, but still in control of himself, and certain now in his own mind that somehow he would—he *had* to—find a way to get himself out. As he strained to look for the thousandth time at the situation behind him, however, he had to acknowledge to himself that he had made no progress whatsoever.

He stopped working. He began to think very hard, as hard as he ever had about anything. He opened his mind to any possibility he could think of; he clutched at straws. One thought he had was that some large animal might come into the ravine—what? a large dog, maybe, or a bear—and if it did, maybe he could get it to come near him, somehow, to dig about the chassis, move or lean against it, maybe, push against the side. He knew this was a foolish way to think, but by now he was willing to think of anything. There was not one single possibility at this moment that he was willing to exclude.

The Holocene

Dogs and wheat were the first of the animals and plants to become domesticated—the dog 14,000 years ago, or earlier, wheat at least 9,000 years ago. Dogs have been eaten by people from time to time and in various places but they are poor food animals. The true food animals began with sheep, which were domesticated some 11,000 years ago, followed after 1,500 years or so by goats and pigs. Peas, barley, and lentils appeared not much later than wheat. The earliest remains of all these animals and plants have been found in the Near East, from the hills of Israel to the mountains of Iraq, suggesting that the processes of domestication began here (although not here alone: the remains of dogs, going back 11,000 years, have been found in Idaho; the remains of squash, going back 9,000 years, in Mexico). Cows and horses are relative latecomers, the earliest cows having been found in Greece, at 8,500 years; the earliest horses, in Central Asia, at 5,000 years.

For the first people beginning to settle down—from 10,000 years ago through the next 5,000, and situated mostly in the southern center of the Eurasian landmass—the menu of available domesticated plants and animals included wheat, barley, onion, lentils, peas, olives, figs, dates, pears, pomegranates, grapes, apples, beef, pork, lamb, and the milk of cows and goats.

The domestication of plants and animals was preceded by the processes of cultivation. Cultivation involves a less active method of manipulation of life than does domestication. Cultivation is a way of harvesting food from animals and plants without substantively changing their nature. Until recently, the Paiute Indians cultivated certain plants through techniques of irrigation which did not require the collection and sowing of seeds. Wild animals can be tamed to the point that they live otherwise than they would on their own. In modern Kenya, for example, the oryx and eland have been coaxed into tended herds. By cultivation animals and plants may be caused to respond to new circumstances imposed from outside their environment without sacrifice of their genetic identity.

But domestication changes their essence. They become biologically different from animals and plants living in a wild state. With the transformation of hunter–gatherers to the new agriculturists, which was marked by the advance in the techniques of food-gathering from cultivation to domestication, some animals and most food plants were changed forever.

The agents of change were themselves changed, subtly at first but in time as irreversibly as the changes they imposed on selected animals and plants. To the people of 10,000 years ago, the immediate advantage of agriculture was that more food could be produced on less land than was required for hunting and gathering. A single hunter–gatherer might cover as much as 20,000 acres in collecting the food to sustain him; a new farmer could live off twenty-five acres. The domestication of animals and plants, which was the beginning of agriculture, made it possible for people to stay in one place—for towns to form, for civilizations to begin. As the numbers of people grew in response to the new abundance of food, more work was required to produce even more food, which produced more people, which required more work to grow more food, which in turn caused some of the people, still hungry, to move on in search of new land to grow more food—a spiraling progression whose inevitable end is just now in sight.

Between animals and plants, the process of domestication is in principle the same. Evolution proceeds through selection that is artificial (in that it is controlled by man) rather than natural. Artificial selection provides desired variations in a plant—from a single species, for example, *Brassica oleracea,* the appropriate pressures have yielded cabbage, cauliflower, Brussels sprouts, kale, and broccoli. Animals of domestic stock can be bred to produce offspring of different shapes and sizes. By genetic manipulation through breeding selection, man modifies and reorders the quality, variety, and quantities of the species that suit him most.

Total domestication is achieved when a species has been so changed through artificial selection that it can no longer exist without the direct intervention of man. Cows untended will not perpetuate themselves; neither will corn. This is true now of many domestic animals and of most domestic plants. At the same time men living from agriculture have domesticated themselves to the point where they are totally dependent upon the animals and plants whose nature they have changed. And whereas the evolution of man goes on now in response to other impulses—to the varieties of social and cultural change—the evolution of domestic animals and plants continues in response to man's choice, subject in their varieties to his trials and errors as against his knowledge, or lack of it, of the processes of natural order.

Santa Barbara, California

Thoughts on the way to see Hardin:

It is the opinion of most of us, when asked to think about it, that the more people there are on earth, the more food ought to be found to feed them. Those who have food ought to try to help those who do not. Whatever one's particular politics nobody wants to see anybody else starve. If it is possible to devise new ways of growing more food to feed those who need it (for example, the Green Revolution), without depriving those who already have it, such steps should be taken.

Most would acknowledge there is at present a population problem aggravating a world food shortage—that there are too many people now for the food available, and that the disparity is growing. The presumed solution to this is something people are vaguely aware of—the suggestion that by helping out the less developed countries, where the problem is greatest, their economies will improve. And as the gross national product of a poor country goes up, its population will go down since people who are better off tend to have fewer children than those who are not. And with fewer people, the demand for food lessens. Meanwhile, and until this happens (as most of us believe), more food should be found to feed those who need it. In such fashion will the imbalance eventually be corrected.

This is the charitable view. In some places it is an ethical or religious view, and in others it is at least a political view. And since it reflects tacitly the announced policies of the major power blocs as well as those of the less developed countries themselves, of the United Nations, whose organizations represent them all—of the haves as well as the have-nots—it is for all practical purposes the world's view.

In every respect, Garrett Hardin of the University of California believes it to be the wrong view. Hardin is an American biologist turned moralist. He derives his precepts from the principles of ecology, which he sees as the basis for viewing the whole of life—its beginning and continuing evolution, including the considerable impact made upon it by man—and as precedent now for salvaging what is left of its future. A biographical sketch describes him as a Republican, a Unitarian, and a grandfather, affiliations which suggest (as does an accompanying photograph) that he doesn't look as radical as he thinks. Rather, with salt-gray sideburns and half-rim plastic glasses, he could be a small-town, middle-aged minister trying to work more closely with the young people in his church.

Hardin was born in Dallas in 1915, grew up at various steps along the Illinois Central Railroad, for which his father worked, and was educated at the University of Chicago where he came under the influence of geologist J. Harlan Betz, the philosopher Mortimer Adler, and the biologist W. C. Allee. He has taught and researched at Stanford and other California universities, and since 1946, at the University of California at Santa Barbara. He is a prolific writer—the author of textbooks, books on ecology for a general audience, extracts, essays, and reviews—and an ardent amateur violinist. The sketch goes on to suggest there is nothing stuffy about him, either. He is a member of the Sal Si Puedas string quartet whose designation translates to "Get out if you can."

But all I know of him is what I've read, and his writing is tough, blunt, and to the point. Will he talk as tough as he writes? One wonders. Although eclectic with his references, often witty and uncompromising with the complexities of biological data, he is almost truculently plainspoken. "A little bit of this kinda stuff goes a long way". . . . "But now the absolute monarch is 'The Pee-pull' "—half wise guy, half buddy-buddy, as though you and I and he know each other well enough now to hunker down and rap. Hardin is a provocateur who writes tough and intimately for a reason—probably, I surmise, to mask the inflammatory nature of his views. Stringing along with his bull shooting you suddenly become aware he has led you into a very grim paradox.

Hardin insists that the *worst* thing that can be done for a starving people who have exhausted their own food supply is to give them more food.

This is a hard line—it is, in fact, a flat-out attack on the most characteristically human of all human attributes: the propensity for sharing, that innate cooperative gesture that is older than history, older than agriculture, older possibly even than human prescience.

Hardin's methods of carrying forward his attack are often as ingenious as they are exasperating. In one of his papers I've marked, "Carrying Capacity as an Ethical Concept," he cites the example of the nineteenth-century algebraist Karl Jacobi, who advised his students, when stuck, to invert the problem; turning it upside down might shake the answer loose.

So, Hardin says, let us take a prominent example of the food/population problem in the particular case of India, a hungry nation of 600 million people living off a decaying land with the carrying capacity for a population of seventy million. The question is, how can we help India? But as evidenced by the fact that we have given food to India and Indians still starve, there has been thus far no satisfactory answer. Now the question in Jacobian fashion is inverted: How then can we *harm* India?

Hardin concedes that gifts intended to increase the sources of energy added to the gift of food might conceivably help such an impoverished nation to help itself—assuming the presence of an infrastructure to ensure proper distribution. But the infrastructure is missing. What can be done about that? Moreover, whereas there has been and may continue to be a surplus of food available in the United States for relief purposes, there is certainly no surplus of energy now, nor is there likely to be within the foreseeable future. "To send food only to a country already populated beyond the carrying capacity of its land," Hardin writes, "is to collaborate in the further destruction of the land and the further impoverishment of its people." That is, so long as the resources are unavailable for helping the country to improve its carrying capacity, the assistance of food will do nothing more than swell the population and thereby accelerate disaster.

He gives as an example elsewhere, with the population clock ticking away: "*Those* (1,000,000) who are hungry are reproducing. We send food to *them* (1,010,000). *Their* lives (1,020,000) are saved. But since the environment is still essentially the same, the next year *they* (1,030,000) ask for more food. We send it to *them* (1,045,000); and the next year *they* ask for still more . . . It is a growing disaster, not a passing state of affairs."

On the other hand: "Fifty years ago India and China were equally miserable, and their future prospects equally bleak. During the past generation we have given India 'help' on a massive scale; China, because of political differences between her and us, has received no 'help' from us and precious little from anybody else. Yet who is better off today? And whose future prospects look brighter? Even after generously discounting the reports of the first starry-eyed Americans to enter China in recent years, it is apparent that China's 900 million are physically better off than India's 600 million.

"All that has come about without an iota of 'help' from us."

Q.E.D. The only way we can help India is beyond our means. A certain way we can harm it is to send it food.

Under the weight of justifying such conclusions it is not his tone alone that seems odd. There are times when he appears willing to try any gimmick to get his point across. In one of his books, *Exploring New Ethics for Survival/The Voyage of the Spaceship Beagle,* he encapsulates his holistic discourses within divided segments of a science-fiction satire. He has, and often takes advantage of, a copywriter's knack for titles ("Guk—Cycle or Sequester?"; "What the Hedgehog Knows"), and he has devised a grab bag of catch phrases in the effort to make his concepts stick:

"You can never do just one thing."

"Everything is connected to everything else."

"Guilty until proven innocent."

"Not to act is to act."

"Thou shalt not violate carrying capacity."

Most of these maxims, deriving from the conjunction of evolutionary biology and human ecology, Hardin calls "pejorisms," a term he has coined to blanket their root assumption, which is that it is more rational to keep the world from getting worse than try to make it better.

Whatever his tactics, however, it is still the horror of unrestrained human population growth that motivates him. I note that in the early forties, after several years spent researching the possibility of converting algae to a large-scale food source—a way to save the starving masses—he concluded that any such measure would only compound the problem. As the human population has come to double during his own lifetime, he has become a dedicated Malthusian, persuaded that uncontrolled population is the ultimate offense committed by man against himself. Recently, such conviction prompted him to say (more temperately than in some of his other writings): "As a scientist I wanted to find a scientific solution; but reason inexorably led me to conclude that the population problem could not possibly be solved without repudiating certain ethical beliefs and altering some of the political and economic arrangements of contemporary society."

While Hardin's ecological views are conventional enough—anti–

industrial pollution, anti–nuclear proliferation, and the rest of it—it is as a polemicist that he is chiefly known, for it is his role, as he sees it, to insist that people now think about the consequences of such things. Thus, in many of the quarrels his work engenders, the argument is not with his science but with the new moral precepts he derives from it. Ecology is a political cause beyond conventional politics. In the early sixties, because it was consistent with his ecological ethic, Hardin became an outspoken and effective advocate of abortion law repeal. And in the early seventies, with what now appears to have been more dramatic instinct than scientific prudence, he began his campaign for the rejection of foreign food aid to overpeopled nations.

Hardin is known best for two essays: "The Tragedy of the Commons" (1968) and "Living on a Lifeboat" (1974). Both are extended metaphors, for it is metaphor, he says, that is the best way to approach unsolved problems. The metaphor of "The Tragedy of the Commons" is built about the choice of farmers in allotting cows to a limited area of pastureland, and it shows Hardin's thinking at its most strikingly effective. From these commonplace references it may be seen how the world's political and economic systems have become vulnerable to ecological disorder, and powerless to avert ultimate disaster. The force of so sweeping an indictment comes from the simplicity with which it is made.

Spare (6,000 words), cleanly written, and free of his occasional slangy abrasions, the success of his essay has drawn little criticism largely because the logic of "The Tragedy of the Commons" is so difficult to fault. Indeed, in a literature moving swiftly now beyond the boundaries of ecology into something altogether new—a sort of biopolitics—it is considered by many to be a small masterpiece. If ecology has become the unexpendable factor in viewing the whole of life, as Hardin maintains, the reception of "The Tragedy of the Commons" would seem to bear him out. In the years since its publication, it has been reprinted in more than thirty anthologies within the fields of political science, sociology, ecology, population, biology, conservation, law, and economics.

"Living on a Lifeboat" is Hardin's attempt to answer the problems raised in "The Tragedy of the Commons." And it is the "lifeboat ethic" that has added infamy to the celebrity of Garrett Hardin. In its bleak reductions of the population/food problem (the metaphor here shifts from land to sea), the "lifeboat ethic" argues that you have a moral responsibility to do pretty much what you must to keep your own boat from sinking, and you let the other fellow take care of himself.

For this aspect of his work, Hardin draws something more than conventional critical opposition. The recurring epithet, in fact, is "obscene." It was so described by, among others, the nutritionist Jean Mayer and the late Margaret Mead. But Hardin shrugs off the word by noting that it is defined as "offensive to accepted standards of decency or modesty," and adding: "Obscenity is, then, a relational term. It defines the relation of an idea to the standards of the speaker. It is an unacceptable relation, so we must always ask which needs changing—the idea or the standards?" However such distinguished enemies may vilify him, Hardin is willing to repeat over and over what he believes to be a new and cruel truth emerging for the human species: "*For posterity's sake, we should never send food to any population that is beyond the realistic carrying capacity of the land.*" It is not the preservation of the present that concerns him so much, he insists, as the rescue of the future.

At the appointed time I am led by Mrs. Hardin past their house to a sundeck within what seems to be a sort of compound, much of it overrun with a laissez-faire mix of garden plants and dry bush—the fuel for seasonal fires which at this moment, to the east, are sweeping the suburbs of Santa Barbara. Pepper plants in wooden boxes are posted at the corners of the deck, and beyond, off to the left, there is a glass-walled building containing a swimming pool. Hardin's study is somewhere in back of that, and after a few minutes he emerges from it, propelling his wheelchair (I had read he suffered polio as a child) toward me, an open, friendly, you-and-I-are-already-pals smile on his face.

Soon enough it becomes impossible not to accept him on such terms. He offers me a pepper fresh off the vine and picks one for himself, and as we munch together, we begin with the food Garrett Hardin has condemned—tuna tainted by mercury, meat and vegetables befouled by pesticides—and the food Garrett Hardin prefers. What does *he* eat?

He takes a bite of pepper and weighs the question. "I'm trying to think of anything that I *don't* eat—that I'm really worried about. At the moment I can't think of anything."

"Not even tuna?"

"Oh, I've changed my view somewhat about that. For a person my age the quantities of mercury are not terribly dangerous. I wouldn't feed it daily to small children—once in a while, maybe. With all these things I think the idea of zero tolerance is a mistake. In general you have higher standards for young people than for old," he says, "because their life is going to go on a lot longer." Still thinking about it, he grins, nibbling his pepper. "I appeared at a congressional hearing on food and population at which I said that as a result of overpopulation we have many losses that most people don't know. A very peculiar moral problem. If you don't miss it, are you missing anything? For example, our standards for sweet corn here are that we don't pick the corn until the water's boiling on the stove. See the corn growing out there on the other side of the orange trees?" He points across the scrub bush, where rising behind the flowers and weeds there are indeed several rows of corn. "I love the taste of fresh vegetables. By the time corn is a day old I think you might as well just throw it out. But the vast majority of people in the U. S. have never in their lives tasted sweet corn. Those poor devils in New York City! The market system simply can't get it to them.

"We're a fruit-oriented family," he goes on. "As for the insecticides, we use virtually none of them in our garden. We figure that's just a tax we have to pay. The insects take theirs, we take ours, and as long as we don't have too much population we can afford to do it. It's when we start grasping, want to get every little bit for ourselves that we begrudge the insects their toll. We have birds peck into our fruit and we try to raise enough so that they can have theirs and we can have ours, too. We like to have the birds around, but they're just hell on wheels on the apricots, damn them!"

Once, at a small dinner party, Hardin found himself seated beside Jean Mayer, one of the scientists who have so scathingly denounced him. Mayer did not recognize him. "I am the obscene Dr. Hardin," Hardin said, and then watched cheerfully as the flustered nutritionist picked over his lunch.

"You are a *Republican?*"

"Yeah." He chuckles easily over this anomaly. "I'd be equally uncomfortable in both parties, frankly, but people identify Republicanism as conservatism, and roughly that's what it is. What I call the true conservative is nearer the ecological position than the true liberal. You know—'Buy hardware! Solve everything! Progress onward, upward!' The real liberal thinks you can merely do one thing. 'We'll pass a law and cure it!' " But as he has so often said, you can't do merely one thing. This is Hardin's favorite "pejorism." In one of his books, he traces it back to a lovely little verse by Francis Thompson, a poet whose life was "sandwiched between Darwin's and [Rachel] Carson's":

All things by immortal power
Near or far
Hiddenly
To each other linked are,
That thou canst not stir a flower
Without troubling of a star.

"The basic concept of ecology is that the world is a vast, complicated, interconnected system," he says. "You can never do merely one thing, so whenever you do, you do a bunch of *other* things. This means that if you're concerned about the total effects of what you do, you carry out ahead of time a technology assessment—you try to predict the secondary, the tertiary effects of what you perform. You have to follow the philosophical principle of 'guilty until proven innocent.' "

"But if everything is connected to everything else, how can you anticipate sufficiently to follow anything through?"

"The answer is that there is no answer. You can never carry out a complete technological assessment, so you do the best you can." That the best will never be quite good enough Hardin affirms in three complementary pejorisms drawn from the laws of thermodynamics:

We can't win.
We are sure to lose.
We can't get out of the game.

It was in 1968, as retiring president of the Pacific division of the American Association for the Advancement of Science, that Hardin first went to work on his now famous metaphor of the commons, in an address he called then "Not Peace but Ecology." Earlier he had found a paper by an obscure nineteenth-century mathematician, William Lloyd, that reduced the concept of carrying capacity to its most appropriate starting point—a community pasture, or "commons." Hardin incorporated Lloyd's perception, adapting it to modern times.

"I had a hard time writing it—I wrote at least seven significantly different versions, because I didn't like my conclusions. I was doing everything I could to escape them. But I couldn't. It was terribly long. Generally a person who's troubled says too much."

He sent the speech to *Science* magazine where "it disappeared into the maw of refereeship. Two months later I got a note—your paper is accepted. It's too long. I realized I had *two* papers, so I took a pair of scissors and cut across the middle. The second part was 'The Tragedy of the Commons.' "

The outcome, the 6,000-word essay which has been now so widely collected, is so tightly written that any attempt to paraphrase it is hazardous. But Hardin, who tends often to restate his ideas in various forms, has paraphrased himself. In "New Ethics for Survival . . ." he briskly sets forward Lloyd's original perception about the commons:

> Very well: *I* am a herdsman using the common. It has a carrying capacity of 100 cows. At the moment it happens to have 100 cows on it, of which 10 happen to be mine. Now I have a chance to acquire an eleventh cow. I wonder if I should do so. I debate the decision with myself in roughly the following way.
>
> If I add one more cow to my herd that will make a total of 101 on the common, which is one too many, and so all of us will lose a bit. On the other hand, 11 cows is 10 percent more than 10 cows, so I gain. The overall loss I would estimate at about 1 percent, and that is shared among all us herdsmen. *My* gain is about 10 percent, and I don't have to share it. Clearly I stand to gain from adding one too many cows to the commons, so I shall do so.

And in a speech on Earth Day in 1970 at the University of Illinois, he went on in even more informal fashion to project the consequences:

> But this same sort of reasoning implies also that [I] should add two more, and three more, and four more, and so on, without limit. The same type of reasoning is used by every other herdsman. The result is that they all add more to their herd, and destroy the commons through overgrazing.
>
> Now the important thing to notice about this is that each man is behaving rationally; each man makes a rational decision when he adds to his herd. . . . And yet the final consequence of all these individually correct decisions is complete destruction of the commons. Since they are all locked into this system, since they cannot avoid this, I speak of their situation as

a *tragedy* of the commons. A tragedy is any course of events which individuals are powerless to prevent so long as they play the game by the rules.

This insight of Lloyd's explains why it is that private property is superior to commons in a crowded world. Under a system of private property where each herdsman owns the pastureland on which his cows graze, the individual is responsible for what happens. He has an *intrinsic* responsibility, because if he makes the wrong decision he's going to suffer from it. He's going to destroy his own pastureland, whereas when the pastureland is shared by all, he does not have that intrinsic responsibility and essentially his responsibility is zero. As a matter of fact, you might say it's worse than that in the commons because he actually has to *gain* by mistreating the commons.

This type of analysis is needed to call our attention to the fact that in terms of their pure forms there are really three basic political systems and not two, as we're usually led to believe. We're usually led to believe that the only choices are private enterprise and communism. But you see, the commons is a system utterly different from either one of these, and is really worse than either.

We need to label this type of a system with a name so that we can speak of it, so that we can discuss it intelligently. And until we do get this label, people are going to continue to suffer from verbal pollution, continue to be misled by improperly used words.

For example, I am sure all of you have seen at some time or another an advertisement showing a fancy, electronically equipped ship plowing through the ocean with all sorts of gear for catching fish. This ad includes a statement of this sort: "Feeding the poor of this world from the inexhaustible wealth of the seas through the benefit of American private enterprise."

Now this is a bunch of crap. In the first place, we don't feed the poor of the world. Only 17 percent of the fish from the oceans go to the poor countries of the world (and even there they probably go to the wealthiest people). Yet the poor countries constitute 67 percent of the world. That's the first falsehood of the statement. The second falsehood is that the wealth of the seas is "inexhaustible." This is not true. We have heard before of the inexhaustible wealth of the American forests, of the inexhaustible game on the American prairies. Every one of the inexhaustibles has been or is being exhausted.

The same thing is true for the oceans of the world. Though at present cropped only to about a third of the extent they could be, the oceans also have limits. The wealth of the oceans is *not* inexhaustible.

The third falsehood is that the harvest is being reaped through the glories

> of American private enterprise. The exploitation of the oceans is not a matter of private enterprise; neither is it communism. Oceans are being treated as a commons. All of us—Americans, Russians, Japanese, Peruvians—everybody who's out there in the ocean getting fish out of it is exhausting the commons, and we are all in a system that has no end except tragedy. There is nothing we can do, each one of us acting rationally, except exhaust the wealth of the oceans. Then we will all be worse off than we are now.
>
> The only thing we can possibly do is to change the rules, and no doubt someday we will. At the present time we don't see how to, principally because we have no international organization to make rules that we will accept. So in the meantime, all of us are playing out this game of the tragedy of the commons in the oceans of the world.

In his more formal presentation Hardin argues that the commons has become something people put things in as well as take things from. The commons is the air we breathe polluted by chemicals and atomic radiation; the world's soil infested by pesticides; the finite spaces overwhelmed by people. "Freedom in a commons," Hardin maintains, "brings ruin to all." How then to curtail this freedom?

"I think it was my daughter's criticism that led me to introduce in the original paper the phrase, 'mutual coercion mutually agreed upon,' " Hardin says. "I was trying to soften the blow. But it bothered me then and it bothers me now. It disturbs other people, too, and I think rightly so."

"Coercion being the forcing of change?"

"Yeah, that's right. As it is now, the family will control the number of children wanted. It turns out on the average that the number is too many. And why should what various families want, by some happy coincidence, turn out to be exactly what the community needs? There's no law of nature that says their perception of their own well-being is exactly consistent with the community's need—there's nothing to *make* it work. So there has to be some sort of pressure. Community pressure. And then the question is, *what* kind of pressure?"

"You're talking about a loss of individual responsibility?"

"I don't like that word," Hardin says, frowning. "I use it in a different way. I say the first thing a person is responsible for is taking care of himself."

"But you start with a man who knows what he has to do to raise his cattle . . ."

"He suffers if he makes the wrong decision," Hardin says.

"Then he has a direct relationship to that suffering."

"He has a direct relationship: that's what he *loses.* Put it another way: if he's one of a group, then his losses are shared by the whole group. But by the definition of the rules of a commons, the gains come entirely to him, and *the losses are shared.* So that's why he's trapped, even though he knows exactly what he's doing is rotten. This is the situation with the whaling industry now. The Russians and the Japanese know what they're doing but they're helpless to stop it."

"But isn't the first problem for them to admit responsibility?"

Hardin thinks for a moment, searching for—what? Another metaphor? "In a book I have coming out I propose the cardinal rule of policy"—it is to be, I sense, another pejorism—"and that is: never ask a person to act against his own self-interest. The most you can ask him to do is to join with you and others in passing a law restricting *everybody.* So it doesn't depend on the individual's conscience. Because if you do that, you reward the conscienceless people. Conscience doesn't work. In order for it to work there has to be an impossibly high standard, one of perfection—*literally everyone* must be ruled by conscience. If even one person in the community follows a lower standard that person prospers at the expense of the others. So the most you can ask is that a law is passed restricting everybody. The freedom of everyone else threatens my freedom; I therefore agree to legislation that restricts the freedom of everyone *including myself.* Mutual coercion mutually agreed upon! In the meantime my conscience doesn't bother me in the least."

"So that if your automobile lacks fuel-emission controls, even so long as there are fuel-emission controls available, you won't put them on your car?"

"I'll *vote* for fuel-emission controls, but I'll not put them on my car unilaterally."

The present troubles began, Hardin has written, two hundred years ago with the laissez-faire ideas of Adam Smith. By looking out for himself, the individual turns out to promote the public interest whether he has intended to or not. The illusion that has come down from this is that progress therefore is good. Present-day industry, producing whatever it can sell without regard to the consequences (to air, soil, water, people) is supported in its destructiveness by this illusion. Technology has become the unwitting enemy, the cost of it accruing in debits of pollution, resource exhaustion, and nuclear hazards.

"I will *not* go out and picket," writes Elizabeth Dodson Gray, as a housewife who describes herself as an ordinary victim of such developments, "but I will attempt to drop out from our technological progress, which I am beginning to perceive is no longer as good for me as I once thought it was. As a friend has said, 'Ours is the first generation with pesticides in our fat, asbestos in our lungs, and radioactivity in our bone marrow.' I am beginning to realize that it is no longer in my self-interest to be a full participant in the 'progress' of this system."

Technology, Barry Commoner has pointed out, has broken free of the cycles of nature. Energy-intensive technology has shifted people from soap to detergents; from natural to synthetic fibers; from wood to plastic; from soil husbandry and land care to fertilizer as the means of increasing agricultural production; and the spread of synthetic chemicals has grown beyond measure. In the West, technology has shifted people toward a new order.

And still another by-product of this shift has come with the widely held assumption that technology can solve any human problem. If there are too many people, technology will find the energy and food to provide for them. If more corn is needed, more will be grown. If the oil runs out, there is nuclear fission, and if that doesn't work, nuclear fusion will.

"The most serious doubts of this view have been put forward by ecologists who have pointed out," Hardin writes, nailing it down again, "that 'we can never do merely one thing,' from which it is deduced that the burden of nonharm should fall on him who proposes to change things, to make an innovation in our way of exploiting nature."

No wonder, then, that the ecology movement arouses such passion —and how scatterbrained it seems to an innocent public when environmentalists try to stop oil producers from disturbing coastal waters. Ecology challenges progress. But it is also a cause for dispute among scientists—life scientists against the scientists of technology. I ask Hardin how the lines are drawn.

"Ecology focuses more on the system," he says. "It is the exponential law of biologists against the inverse-square law of physicists. Physicists who were until recently seen as highest in the pecking order of scientists made progress by simplifying systems—by eliminating the insignificant. But to the biologist, nothing within a system is insignificant.

"Consider the word 'exponential' with its overpowering implications," he has written elsewhere. A single bacterial cell lands on the lining of a man's throat. Does it matter?

"A Newtonian physicist, after determining that the bacterium weighs only 10^{-18} times as much as the man—only 0.000000000000000001 as much—would dismiss the event as inconsequential.

"A biologist would not. He would ask first, 'Is it alive?' If the bacterium is alive and capable of reproducing on the lining of the throat, the population of its descendants may double every twenty minutes. In four hours there may be 4,096 cells; in eight hours 16,777,216; and in a day . . . well, in a day, if the cells could keep up so brisk an exponential rate (which they cannot) there would be more than 10^{21} bacterial cells in the man, and the aggregate weight of them would be five thousand times that of the host. (Now we see why the exponential rate of increase cannot be maintained.) All this from only one cell . . . increasing exponentially."

All well and good. But are such sensitivities sufficient to form an ethic? And so harsh an ethic as Hardin proposes? If man is a moral animal, what can be moral about an ecological ethic? His answer lies in the evidence derived from the systems of living things—in the behavior of populations of any species. Food encourages reproduction until all the food is gone—but once it is *all* gone, when the capacity of the source to renew it is exhausted, it is not only the starving that lose, but the future as well. The essence of the ecological ethic, he asserts, is that it pays attention to posterity—"to our grandchildren," he has urged, along with Frankel—"everyone's grandchildren." Wittgenstein says there are some responsibilities men have beyond the grave. These Hardin seems to have taken as his concerns. He is far from the dour doomsday prophet much of his work would suggest—nor is he, he says, hopelessly pessimistic about the possibility of corrective action from an aroused, and informed, public opinion. He cites the shift of attitude in this country toward antiabortion laws as an example of a revolutionary turn in social and religious conventions occurring within only a very few years. It is not too late for things to change.

He is still working out the lifeboat ethic, but he is less sure of his ground here. What I must understand—must have a very clear understanding of to follow the lifeboat ethic—is the meaning of carrying capacity. Hugh Lamprey believes it to be a self-evident term; the others assume it is clear enough. Hardin is at some pains nevertheless to describe it to me now in detail:

"Carrying capacity is defined as the maximum number of a species that can live in an environment without degradation of the quality of life for that species indefinitely into the future. For most species it's so hard to know what the total life style is, it is difficult to measure. You have to make allowances for the fluctuations in the environment, so that you allow a safety factor by calculating that the actual number is always somewhat less than the apparent maximum. But you can *deduce* the carrying capacity inversely by looking at the environment."

Do I follow this? I am not sure that I do.

"If the environment is becoming degraded," he goes on, "you can assume the species has exceeded carrying capacity. For example, the deer in Wisconsin. If you find that the trees are being eaten higher and higher by deer in the winter time, and the young trees are dying because of excessive browsing, then you can say the carrying capacity has been exceeded. So we had better kill some of the deer. It's hard to determine it directly, but with that qualification—measuring inversely—you can establish the carrying capacity with a reasonable degree of accuracy for all animals—except man.

"Here the complication is that for the past two or three hundred years, technology has been increasing the carrying capacity of the environment, which has caused us to pooh-pooh the idea of carrying capacity altogether. It *is* obscured, of course, and this is the dreadful thing about technology. Over here—in the West—we have been so good at it that the rate of agricultural improvement has been 3 or 4 percent a year for the last fifty years. Fantastic! But it would take only a little shift to cause problems that could become exponential. Look at India, for example, where the land has been so badly treated. Every tree along the road south of Madras has a scalped place on it, with a license number, to make sure it isn't cut down. When nightfall comes, people snip off twigs for fuel. In Kashmir they are making charcoal out of leaves. It makes piss-poor charcoal but that's what people do at the last, when they're grabbing onto anything."

Because it is Garrett Hardin who has raised the comparison of human populations to other animal species, and because it is this question which has brought me here, I ask him as I have the others:

How can it be that we as animals ourselves, evolving out of the animal kingdom, have lost the capacity to adjust to environment?

"There is nothing in any other animal species that corresponds to policy," he says. "So there can be no concern with death. So you have your wildebeest, which suffer periodically such large losses, and so what the hell? They have no power to control their numbers. And for most species, most of the time, the deaths that occur come as a result of predation either by macropredators—lions or tigers, for example —or by micropredators—disease organisms. We've created *our* problem because we've gotten rid of almost all of our predators."

"But this is only recently, isn't it?" I reply. "We have been in balance until our numbers took off. You've written that if the human world was a smaller sphere there would be no problems—with technology or anything else. Within the natural laws of other forms there is not a conscious or willful ruling of how many there are to be—things are worked out by the laws of nature rather than of consciousness. So mustn't you say that man has moved away from the laws of nature in this respect?"

"Only in his refusal to substitute conscious, human *birth* control for the death control he's gotten rid of," Hardin says. Mrs. Hardin arrives at the sundeck with a pitcher of ice water, and I welcome the break. I wonder if I am simply asking the same question in different ways. I decide to start over.

"Richard Leakey says it is a species failure . . ."

"To *date* it is a species failure. I don't think it is a necessary one, nor that it has to continue. It's more true of our part of the world than any other. For us anybody who prevents a human death is regarded as deserving a first-class hero's medal. This isn't true of all cultures; many regard death as far more of a friend. With us it is very strongly a Christian-Judeo thing, I think, and maybe the two together make it worse than either alone. Tertullian, in the third century, one of the principal founders of Catholic philosophy, believed that disease and war were natural things: we need them to have a lopping off of the superfluous branches of humanity. We no longer believe that we need disease and war, but Tertullian did, and he was a Christian."

How, then, as an evolutionist, does he feel about the immediate future of our species? As bleak as Evans?

"I think we're in a momentary situation of temporary insanity. We seem to think that the species can be saved only if every individual life of every member of the species is saved. And when you couple that now with the 'right' that is being expressed in the United Nations—the 'right' for each family to determine the number of children that it wants—when you put those two together you have just an inflammatory mixture."

"So all of it in your view comes down to this disregard of carrying capacity?"—a cue for him to continue into his ethic which he takes up, seemingly not disturbed in the slightest by the overtones of what he has just said.

"But another thing enters with the carrying capacity for humans. You have to define it in terms of the quality of life you want to live. An easy example to take is of eating meat versus not eating meat. Because of the ecological 'tithe' as we call it, ten pounds of vegetable matter are required to produce one pound of animal matter. If you want to eat meat, the carrying capacity of your environment is not as great. So we have to ask, carrying capacity under what standard of living?

"And *that's* part of the problem. Very quickly you get into an argument about values. Automobiles, radios, TVs, and so on. If you didn't have such things you might carry a few more people in your environment.

"So a very real question is: is God giving prizes for the maximum number of people? Most people, unfortunately, act as if He is—as if it is our purpose on earth to produce the maximum number of people. For what end? Nobody explains why it's better to have ten billion people than two billion . . ."

"So then it is population which concerns you most?"

"*Yes.* Because increased population growth at the present level of human existence increases the magnitude and keenness of *all* the other problems. That *wouldn't* have been true two hundred years ago, but we've now reached the point where I don't know of a single thing we could do that we could do better if we had more people. Whereas I know many things that we would do worse. Such as having fresh corn."

"So if your ethic is predicated on there being too many people and too little food, what do you see as the solution?"

Again, he corrects me. "I wouldn't merely put it at too little food. But too little resources of all sorts—food, energy, amenities. Which brings us back to the problem of responsibility and power. If a nation asserts it has sovereignty over its own affairs, including its own reproduction, then the rest of the nations should insist that sovereignty carries with it responsibility for survival. In other words, if the people of a nation produce too many of themselves for the carrying capacity of their own land, living the kind of life they want to live, and then appeal to us, say, to bail them out, to send them food, send them energy, we should just simply ignore them. Sovereignty means *responsibility.*"

Like Frankel, Hardin raises the example of China as an ecological model although he professes to know very little more than what he has read about it in the press. Instead of national coercion in forcing people to conform to their environment, there is community pressure. Within a commune of up to 30,000 people is the squadron of a few hundred, which is the basic unit of responsibility governing the lives of its members. He has little interest in ideological differences; Russia and the U. S., for example, are equally backward in ecological awareness. The diversity of cultures is something he favors—"the optimum condition for the survival of the species as I see it. It would be better if we were even more separate than we are now. Many people want us to make one big culture worldwide. We're not wise enough. Let's have many, pursuing different value systems, and let some of them fail—an object lesson to the others. The greatest mistake we could make, I think, would be to adopt the idea of an universal declaration of human rights. That can be reciprocal. The ideas of others are different from ours. Since you can't be sure you have the right answer, you should be a little bit humble about such things."

"But what then of the problem of introducing birth control into countries whose cultures are built about large families? The population problem is in the Third World."

"You can't do it. All we can say is, here are the external conditions you'll have to meet. No handouts. If your people are starving, we'll do nothing to save them. We should give *information,* of course, as much as is desired. We could say, by the way, we happen to know some rather neat methods of contraception, in case you'd like to use them, but it's still *your* problem."

If Garrett Hardin were president, what measures would he take to meet the crises he sees forming?

"I guess the principal thing I would do would be to try to set a tone of individual and of national responsibility. No nation should suggest any other nation has to pull its chestnuts out of the fire. Immediately people say that's self-serving—we're so rich we don't need any help.

But I would point out that with respect to oil we're not rich; we are importing up to 45 percent of it. Other people have the problem of starvation and we have the problem of energy starvation. And we've *got* to solve it because we can't expect other people to solve it for us. I'd return to an old-fashioned sort of morality—as Kennedy did when he said, 'Ask not what your country can do for you, ask what you can do for your country.' I'd return to the idea that we should be *self-reliant.* This is what Mao did for China in 1948. And that means educating the public to the difference between self-reliance and self-sufficiency. We haven't a *chance* of being self-sufficient. We require minerals from many parts of the world. But we *can* become self-reliant. If we need chromium, we trade wheat for it—we don't ask for it free."

"But don't you put yourself in the position of being boycotted?"

"Indeed yes! We should not expect any gifts. I'm talking about a revolution in thinking. *Yes,* this is contrary to the growing tendency of the past hundred years of the One World idea, the brotherhood of man, and so on. I don't think we can afford such ideas. We have to have smaller groups, responsible groups, sovereign groups, powerful groups, groups able to make mistakes, to suffer and learn from them. And *not* get saved! And that includes us."

Hardin's use of the lifeboat metaphor to support his arguments is simple enough. The carrying capacity is marked on the bow—forty-four, say, for safety; fifty, maximum capacity. In extending the metaphor, he applies it to nations in terms of their relative wealth. "Approximately two-thirds of the world is desperately poor, and only one-third is comparatively rich. The people in poor countries have an average per capita GNP of about two hundred dollars per year; the rich, of about three thousand dollars. (For the United States, it is nearly five thousand.) Metaphorically, each rich nation amounts to a lifeboat full of comparatively rich people. The poor are in other, much more crowded lifeboats. Continuously, so to speak, the poor fall out of their lifeboats and swim for a while in the water outside, hoping to be admitted to a rich lifeboat, or to benefit from the 'goodies' on board. What should the passengers on a rich lifeboat do? This is the central problem of the 'ethics of a lifeboat.' "

While the exact limits of the carrying capacity of any nation may be unclear, certain shortages enable us today to see that *most* carrying capacities are being exceeded. For ourselves with oil, as he says, it is clear that we no longer have enough. "We have been living on capital," Hardin warns, "in the form of stored petroleum and coal, and soon we must live on income alone." With food, a rampant plant disease may destroy crops; we are obliged to maintain some reserve as a safety factor.

"The fifty of us in the lifeboat see a hundred others swimming in the water outside, asking for admission to the boat, or for handouts," Hardin writes. "How shall we respond to their calls?" The only legal precedent has to do with actual lifeboat situations, the law excusing those aboard who, to survive themselves, refuse to take aboard the drowning. For those who feel this may be unjust, Hardin adds coldly, "The reply . . . is simple: *'Get out and yield your place to others.'* "

"But can you then extend your lifeboat ethic to suggest this is what every nation should do?" I asked. "Don't you presuppose parity of some kind among all the nations—that everybody has something that everybody else wants?"

"No. The carrying capacity is the property of the environment, *not of people.* The Galápagos Islands have something like five hundred people on them. They shouldn't have two million. If a place has too many people, it has got to cut down. If any country wants to accept some immigrants from Tanzania with its fourteen million, fine. But if not, that's Tanzania's problem. Each country has to regard its carrying capacity as a fixed limit that it has to adjust to. That is its responsibility."

I thought of Frankel's reference to Schumacher's *Small Is Beautiful.* I asked Hardin if that wasn't the way he seemed to be heading.

"Schumacher and I don't share fundamental views, but very important secondary ones. He is a voice of sanity, with that marvelous phrase, 'appropriate technology. . .' " But there is no point in going in that direction now, Hardin's "ethic" still hanging in the air, forcing its impossible terms.

Beyond its logic, the reality of Hardin's way of arguing for human intelligence (birth control) over animal fate (death control) remains harsh to the point of sickening horror—perhaps all the more so because the man before me, moralizing congenially in the balm of the California sun, has accepted so resolutely the consequences of his own analysis. Or perhaps it is because I am just unable to accept his version of the truth he urges, or that his argument under present circumstances is indeed the only humane one to be made. Without prompting from me, he continues:

"I think our only chance of survival is to say it is quite within the rights of any nation to refuse to take aboard other people. The fact that they are starving has nothing to do with it. When all those people were starving in the Sahel several years ago, many suggested that the neighboring countries should take some of them in, but they didn't suggest it in a very loud voice. Because if they did, they knew the question would be: well, why don't *you?* Nobody did, and I think it was recognized then that no neighboring country had such an obligation. A sovereign country is responsible for its own survival. I think this is the best approach, and a humane approach."

I quoted some of his writing back to him: "You say, in three different places, that for posterity's sake we should never send food to any population that is beyond the realistic carrying capacity of its land; you say that food shouldn't be given without attendant population control measures guaranteed; and you say that food shouldn't be given without supporting grants of energy. So what you say"—and I came down hard on it to be certain I had his position clear—"is in effect that there is not going to be any food given to anyone."

"In practice it would usually turn out that way," he replied. "Though as a matter of principle you should insist that we would like to help in all three of those ways."

"But what would it do to the people of the West, within the traditions of humanitarian values that have come out of our own cultural evolution, to make such a judgment when other people are literally starving all over the world? Say that they do *not* perceive such action in the benign terms you argue—that it's for the sake of the people of the future—but they simply accept it as policy. What's all this got to do with me, one might wonder, turning on his television set to watch it. One is either going to resent the policy terribly and feel the guilt of inaction, or simply become indifferent. In either case isn't this the equivalent of genocide?"

He doesn't flinch, but he doesn't look very happy either. "I think that's a good criticism," he says. "On the intellectual side, I think the carrying capacity analysis is a correct one. But in terms of *selling* people on the realities . . ." It is an odd verb to use. "Here I'm not a very good guide. I'm too far from the position of the rural huckster —the person who really sells things to the public. I don't have a keen enough instinct. Probably, if someone came along who could do it, it would be done by means I disapprove of, that I regard as blatant, cruel, bad form. I don't know what will happen . . ." He is quiet for a moment. "And this does disturb me. I'm trying to keep it on an intellectual level."

13

GETTING OUT

Trout Lake, Washington

He went over everything in his warehouse, again and again. He fingered the wires, the tie rods, the tennis racket, the suitcase, the fabrics, ran his fingers under the dash. He hefted the tire iron. He made himself think about them all in different ways. Had he overlooked any possibility? Did he have something he didn't know he had? Should he try to think of new ways to use the same things?

He turned back to the tire iron, and looking at it again, at its beveled front edge and its socket opening at the end, he thought about this tool. It was designed to pry loose hubcaps and unscrew lug bolts. The edge was beveled. The edge was beveled. The edge was *beveled!* The beveled edge of this tire iron could be a chisel. With a chisel—if he had a hammer—he could cut through wood. But of course he had no hammer. But a hammer was nothing more than a shaped weight. What did he have that would make a shaped weight? He looked through the warehouse. Nothing would serve this purpose.

A rock could work as a weight, one properly shaped with a good heft to it. There were plenty of rocks out beyond the car. (Didn't he know it?—they had snagged his line often enough.) If he could somehow fish in a large enough rock of the right dimensions, he would have a shaped weight, and when he had a shaped weight he had a hammer to go with his chisel, and with the hammer and the chisel he could cut through the wood root that was holding his foot.

Vihtelic could barely contain his excitement at his discovery. He reached as far as he could—it was dark now—under the chassis, clawing through briars and dirt for a rock that might be hidden there. Nothing. Calm down now, he told himself. This would have to wait until tomorrow; and he did as he thought he should, wrapping his sleeping bag around himself and working through the night at sleeping as much as possible.

Very early on Sunday he concluded the only way he could get a rock of the size he needed would be by scraping it in, and the only object he might accomplish this with was his suitcase. By tossing the open case out beyond the car, he could eventually aim for the rock he wanted; the problem then was to get the suitcase back. Cautiously he threw it out a few feet, and then, bending a hook into the end of his tie rod, he hooked into the handle of the suitcase and dragged it back. That was okay. So—there was no little risk to it, one toss too far and goodbye suitcase—he picked his target and aimed for gold. It took him most of the day to perfect the system, but well before dark, the suitcase scrabbling his hammer toward him in response to the delicate, gingerly pull of the hook, Vihtelic held in his hand a heavy, flat-headed rock. Without wasting a second, he placed the beveled edge of the tire iron four inches above the point at which the root held his foot. Reaching behind him with the rock, he struck the near end of the tire iron as hard as he could, and only once. *TCHUNK!* The edge of the tire iron went so deeply into the root he was at first unable to wrench it out.

Vihtelic was as comfortable now, as secure, as self-contained as he had been anywhere else in his life, under the most conventional of circumstances. The reason for this was simple and obvious to him. Now he could get out when he wanted to. It was not practical to do so in the dark; so he would wait until first light tomorrow. What was the rush? No problem. Calmly, almost serenely, he wrapped himself in his sleeping bag again and went to sleep.

On Monday morning, his sixteenth day in the hole, Vihtelic took his chisel and hammer and prepared to cut into the root that held him. But wait a minute. The car was balanced on his foot and the root under it, which together acted as a fulcrum. Once the root was cut through, the weight of the car could simply drop farther, with his foot still caught beneath it. He looked for something to serve as a wedge, to support the weight of the car once the root under his foot was cut through. He was anxious to get going but he wasn't going to be foolish about this. Not now.

He found a heavy root, slightly smaller than the one he had to cut. He wedged it under the chassis of the car as far as it would go, and then he took his chisel and his hammer and began to cut into the root that held him, one chip at a time, and as the pieces began reluctantly to give to the bite of the iron, one falling atop the next, slowly, with long, painful intervals between them, one falling and then bang, bang, bang, no matter how it hurt to turn back on his dead foot, wrenching his body, turning and banging, the small chips falling away and the size of the root growing smaller, bang! bang! Vihtelic knew that very soon now he would be free. . . .

The Holocene

Over the last 40,000 years, men have got where they are now not by new stages of biological evolution through natural selection but through cultural evolution. Whereas their physical characteristics have remained the same through these years, they have learned to change things, themselves, and the environments in which they live. Learning is the selection process of cultural evolution: what doesn't work dies out, what does work is continued. (Some things that don't work continue too, but possibly this is because the selection processes of cultural evolution, as in biological evolution, proceed along different levels of time.)

Animals change genetically through natural selection to adapt to their environment. Through the genetic inheritance of the ability to learn, men change the environment to suit their needs. Through natural selection birds have evolved wings which enable them to fly; through learning, men *make* things to fly themselves where they want to go, even beyond the earth. Whereas animals rely heavily on instinct to sustain them, modern men depend so completely on the accrued learning available to them as to be nearly helpless without it. Indeed, it has been suggested that should a generation of them be deprived of it the succeeding generation would revert to the Stone Age.

But that is to force the point. The hold of men on their cultural heritage is fragile enough without an interruption of generations. In modern times the Ik in Uganda, forced from ancestral hunting grounds into an environment from which they could not derive food, moved through starvation toward the abandonment of their own children; in one of the most civilized of nations, the Nazi Party incorporated as political policy the systematic destruction of twelve million people. Learning alone has not provided insurance against social deterioration, but it is the only way man has of advancing his interests.

The human baby is notably helpless at birth. Except for motor reflexes and very little else, it comes to function by learning, and much of what it learns—through thinking—comes to seem instinctual. The methodical, ordinary, workaday behavior of most people carrying out ordinary things—writing a letter, driving a car, reading the paper, even much of the work they do for a livelihood—is the consequence of considerable thinking by others going before, but these activities require little or no active thinking now to carry out. (Learning to think about things is another matter. It is the only way to the act of creative imagination, which is the highest form of thought.)

Beginning slowly, the thinking processes in early humans came in later years to accelerate very swiftly, so swiftly they are now seen to have progressed in evolutionary terms as "saltations," or discontinuous variations—the dramatic transformation of one form into another: from tools to fire, to shelter, agriculture, writing, steam, and then combustible engines, to television, space vehicles, and nuclear power (the last of these within the present century!). Things thought out for one purpose open new and unexpected paths. The first electronic computer was designed to collect ballistic data for artillery during World War II. Now the computer extends the range of human thinking: it amplifies intelligence. A sophisticated mathematician would need a thousand years to make the calculations a computer can manage in hours. The computer can not only trace the birth of a star but can calculate the varieties of consequence of human social actions, thereby increasing the possibilities of choice. In service to the human brain, the computer is yet another saltation in the cultural evolution of man.

As the center of civilizations has shifted from the Old World to the New, the assumption has inevitably grown that man can do anything he wants, or has to, through the practical applications of his ability to think, i.e., through technology. Through agricultural technology he so modifies his environment as to assume a limitless abundance of food. Through medical technology he has blocked diseases which only seventy-five years ago could be fatal—scarlet fever, malaria, diphtheria, typhoid, smallpox—and thus assumes a victory soon to come over all disease. Through industrial technology he has created mechanical devices which extend his senses to the bottom of the sea, into the heart of a cell, outward into the universe; and made instruments of defense —there are 13,000 nuclear bombs stored on the European continent —as awesome in their complexity as they are lethal in their force. He gains in his ability to manipulate atoms and genes, the constituents of existence. . . .

Caterham, England

Niko Tinbergen, the Nobel Laureate ethologist who works mostly with seagulls, is acutely alarmed over the accumulation of saltations in the latter half of this century—the vaulting of men beyond natural limits to the state he calls "disadaptation." Of such magnitude are the problems ahead, he says, a corrective change, "even if we were to start now, would require more time than we have available." It is not just alone the crisis brought on by the energy running out, or the pollution of the atmosphere, or the disturbance of the climate, or the exhaustion of mineral resources, or the social disarray in an overpopulated world overloaded with nuclear destructive capability, but it is all of these things—a biological crisis for a very mortal species. They are all interrelated, each as tightly bound to the other as hydrogen to the oxygen in water.

Little I have heard or learned fails to bear him out. Moreover, and even more terrifying, if that is possible, such problems have not only moved beyond conventional political remedy, they are as yet scarcely perceived by the world's political leadership. Is it because they are so deeply rooted in time? And that they therefore seem to be only fragments of another crisis in a century of crises which, like those having gone before, eventually will pass?

Or is it that the past of recent memory is in other ways too reassuring for alarm? Science and technology deriving from Western civilization have fostered the illusion that habitat can be domesticated at will to the service of humankind. Food supplies multiply, disease dwindles, and industry stimulates appetites beyond need. Less than a century ago, there was concern that there were too *few* people. People in the West saw food surpluses burned to keep prices stable. Energy sources have been plentiful. Through much of the nineteenth century the dominant source in the West was wood, which, before it was depleted, was replaced by coal. Before the coal was depleted, oil replaced coal. The illusion has been that something will always come along to fill the need before the need becomes acute—the technological "fix." Although the most optimistic review of the past does not bear this out, the illusion nevertheless persists.

With the advent of the computer, the Faustian choice has been sharpened, for now there are foreseeable consequences with measurable variables. For example, five times more than the known available quantity of fossil fuels would be needed to sustain the rate of consumption in America of the past ten years for ten years more, and then they would be gone. (In each of the energy source exchanges of the American past—from wood to coal and coal to oil—a period of forty to sixty years has been required to bring each new energy source to the point that it would provide 10 percent of the nation's needs.) Agricultural land in America is shrinking at the rate of three million acres a year. The deserts in Africa are *growing* at the rate of three million acres a year. The world's fresh water (less than 1 percent is available as drinking water) is being polluted at the rate of 2 percent a year. The rain forests are shrinking at the rate of fifty acres a minute. New chemical substances, a thousand more each year than the year before, are entering the biosphere, with unknown consequences. When the course of this century is completed, the people of the earth will have quadrupled in number. In ten years, at the mid-seventies' rate of consumption, much of the available mercury and silver will be gone; in twelve years, much of the tin; in fifteen, much of the zinc; in eighteen, much of the copper—and so on and on.

In *The Ascent of Man,* Jacob Bronowski said that no belief can stand in this century that is not "based on science as the recognition of the uniqueness of man, and a pride in his gifts and works." Nevertheless, Tinbergen warns: "One single lethal negative feedback can undo all the gains of the cultural evolution, and can even threaten our very survival."

Schumacher is the last on my list, an economist! The choice, like the directions of my travels, had been made for me. He was the one I should see, others had said; any holistic solution would necessarily be suspect, but he at least had tried to address such problems.

It is about nine-thirty now, Thursday morning, my train several miles out of London and proceeding in a southwesterly direction. A sooty industrial ghetto slides past my window. I crammed his little book, *Small Is Beautiful,* in my hotel room last night, and flipping through the clippings I have about him, I have had to guess that as economists of the West go, E. F. Schumacher would seem to approach the lunatic fringe. Among other things, he is against what most of them favor, an expanding economy—or, in the jargon of their science, "capital-intensive, high-energy-input-dependent, human-labor-saving technology," whose aim is to increase the gross national product. "More, further, quicker, and richer," as one European economist neatly puts it, is the objective Schumacher sees as inherently violent, damaging to the ecology wherever it is applied, heedless of finite resources, stultifying to the individual and indifferent to limitations.

The clippings say his views began to form twenty-five years ago, when he served as economics adviser to the prime minister of Burma. He became intrigued by what he describes now as an inverse relationship between technology and human well-being. The less industry, the more harmony with place and one's fellows; the more leisure. (I remembered Frankel quoting Hardin on the golden age of the early hunter–gatherers.) As far as production was concerned, Gandhi seemed to have the best answer: "The poor of the world cannot be helped by mass production, only by production by the masses." Work was a *value,* not a chore; a right which not only sustained the individual but in fact ennobled him.

Since so much of this philosophy does not derive from the spirit which built the modern West, Schumacher is not exactly the sort *Fortune* magazine, say, would solicit for countertrends to the inflationary spiral. Nevertheless, his credentials are imposing: a Rhodes scholar from Germany, a lecturer at Columbia University; a protégé of John Maynard Keynes; lead economics writer for *The Times* of London; economic adviser to the British Control Commission in postwar Germany; and chief economist for the British Coal Board (where he worked with the more optimistic Bronowski). Since his retirement from the last, he has consulted independently with the forming governments of many Third World nations, including Tanzania, where Julius Nyerere is said to have borrowed from his ideas in composing his Arusha Declaration—"Ujamaa," a forward-looking African Socialist scheme to return the people to an agrarian life compatible with their traditional culture. But mainly it is his little book that gives his ideas currency at the moment. "You want to know my philosophy?" Governor Jerry Brown of California asked at a press conference, holding it up. "Here it is." Scarcely reviewed upon publication in 1973, it has since been translated into fifteen languages and received the ultimate canonization of being added to the curriculum of hundreds of Western universities.

Small Is Beautiful is certainly more widely known than Hardin's "Tragedy of the Commons," and although in a sense it treats many of the same problems, it is not just Schumacher's analysis of environmental catastrophe that sets it apart. Nor does it provide an especially novel critique for the causes of such problems. Political remedy is scarcely addressed, and many of his assumptions—among them that bigness is necessarily bad—are open to question. As the twentieth century and the second millennium A.D. rocket hell-bent toward their close, however, and with the doomsday chorus swelling, it is for many the appeal of E. F. Schumacher's book that what he is after, given our present difficult circumstances, is a low-level, pragmatic, inexpensive, do-it-yourself program of survival for the species.

Since we don't know each other by appearance or otherwise, I am to proceed to the station car park where he will be waiting in a green Volvo. There, a large man I guess to be in his mid-sixties unwinds himself from the driver's seat to greet me. He has long straight gray hair he combs back from his forehead, scruffs of it continuing past his collar in back; and he has long gray muttonchops. There is an oddly wizened look to his face; the lips are pursed. A child of about four is playing in the back seat. "Your grandson?" I say, with a friendly smile. "My son," he says without expression, and I push my smile a little harder. In silence we pass through the town and into the country. His house, set into a wooded hillside, is modest, neat and spare; a workman's muddy rubber boots wait in the front hall. The success of his book and his great popularity on the lecture circuit would have enabled him to enjoy something more elaborate, but I have read that Schumacher is as good as his written word, living simply, traveling light, and devoting himself vigorously to the needs of others. It is almost the house of an ascetic, and of a strange ascetic indeed, I reflect, who seeks to reconcile the disparate problems of Julius Nyerere's Tanzania and Jerry Brown's California.

It is, I suppose, one more signal of how far I have come that I find myself genuinely startled to discover that E. F. Schumacher believes in God. In fact, for the first five minutes of our conversation—a monologue I could not have interrupted if I tried—I found I was in for a diatribe directed at correcting the import, if not the details, of most of what I have heard in the three weeks preceding my arrival here. "Let's get right into it," he said. (I had written ahead the scope of my interests.) "Your time is limited. I think our civilization is being *torn apart* by the conflict between science and religion. In *such* a manner *that* when you try to talk about ethics, or right and wrong" —here, with an exaggerated whine, he mimics what I assume to be his adversaries—" '*That's* not scientific! That's mumbo *jumbo!* That's *any*body's opinion! When you pursue science, you must do so in a totally unethical *way!*'

"This, I think, is the *cancer* we are suffering from. For the past three hundred years we have *mesmerized* ourselves with the belief that we understand the universe as a physical machine, that there is really nothing sensible that can be said about right and wrong, and how we ought to conduct our lives. 'It's just anybody's guess!' And the main support of this terrible and terminal cancer—if we don't *do* something about it—is this . . . *theory of evolution!*" The indictment ends with a hiss.

Before I have quite absorbed his words, he broadens the attack. "The idea that everything just starts with a Big Bang, and there is no *mean*ing, no purpose at *all.* That through an accidental convocation of atoms you and I result, and that the whole thing is just as *stupid* as anything could be!" His foreign accent is slight but his speech rhythms are still vehemently Germanic and his rhetoric so ensnarled as he switches from heavy sarcasm to the earnestness of sweet reason that it isn't easy to tell who is supposed to be speaking. One must listen for the quotes within quotes. He is, in any case, certainly overwrought. " '*There's* no intelligence behind it!' "—the adversary speaking—" 'No *spirit* behind it! The spirit, the intelligence are epiphenomenal! The human race is a rather sorrowful accident of evolution in the universe. Maybe on other stars similar disasters have happened. There *really* isn't any difference between animals and man!' " He shrugs.

"The result of this is that any thoughtful person, after he or she has absorbed this intellectually, will come to the conclusion that there is *no* meaning. The *whole* of evolution shows that there is only one principle in the universe and that is—utilitarianism. *Survival. Natural selection!*" The adversary again: " 'I mean, *all* this is perfectly all right. And *there* is no meaning. *There is no responsibility!*' " I realize I am becoming an audience for him—not the cause of his ire but a convenient occasion for it.

"All of this," he goes on—Schumacher now speaking for himself

—"is a kind of *hoax.* It is *not* science. It is a *faith,* and it happens to be an extremely degraded faith, the degraded faith of utilitarianism. We can't live without some kind of faith, and we may even adopt this one in spite of the evidence revealed—that the whole of nature is beauty through and through. *Certainly* it is not something to be explained as a utilitarian machine! All of this *odious* stuff about the Big Bang! And these useless people like Fred Hoyle who get into the papers by saying, 'Oh no, I've changed my mind. It wasn't *four* billion years ago, it was *five* billion years ago!' Or those earth-grubbers in Nairobi who dig up some old bones and launch their theories! *All at a time when we don't know how to run New York City!*"

Was this man a Creationist? Was I in the right room? The right century? I snatched at the opening: "What are *your* views on the origin of man?"

"Speaking as a scientist there is only one honest answer. *I don't know*"—he bit off the words and stopped, staring intently at me. "Because there is *no* scientific knowledge that enables me to explain such things. The allegedly scientific knowledge of biological changes, of evolutionary theory, is so trivial compared with the size of the question! I'm a plant breeder, I know about these things, but I also know jolly well that I have absolutely no possibility of ever breeding into a plant the kind of intelligence that sits opposite me now. To say that this simply happened—*that* is the mumbo jumbo! As a scientist I have to say, *I don't know.* It is one of the vices of the modern mentality that we are not frank and honest enough to *say* we don't know. We are too conceited. But as a Christian I say, '*Now, wait a minute!* The mainstay of understanding that I have felt in my life is revelation. And I have verified that it is the same in all the great religions. I have lived with Buddhist monks in Burma, and I have studied Islam, and so I must take revelation seriously. If you take the Chinese language—and within it there is talk about God the creator, that's good enough for *my* purpose—which is to live! To make my way through life!'

"Study the things of how to *live,* and what's *right,* and what's *wrong!* That is more subtle than all this, this . . . *natural science stuff.* Thomas Aquinas said, 'The slenderest knowledge of the highest things is more desirable than the most precise knowledge of lower things.' But during the last three hundred years, as epitomized by Descartes, we have said, 'Only where you have the most *precise* knowledge is it science at all!' So we limit ourselves to lower things"—and he is off on it again—"like weird and wonderful theories of a Big Bang. I mean, even the *language* of it is so odious: *'the original soup!'* And that's so *stupid,* because nobody can tell me where this soup comes from.

"Our forefathers said, 'Let's not talk about meaningless things—whether the Lord started this whole thing with a soup, or with a bang; whether he took a short time or a long time over it. All this is *completely irrelevant.*' And it's only in a state of degraded materialism, that is *killing* us, that these things receive *any* attention. Tibetans, for example, just laugh about it. 'You white people, you Western people, are just *children.* You haven't grown *up* yet! That's the very thing you have to grow *out* of!'

"We only know things that we can do ourselves. Somehow the doing must come before the knowing. And we haven't made the world. We *can't* know. We can only contemplate it, and that is a very worthwhile thing to do. As a religious person, I say *yes,* I can accept in a naïve sort of way that there is a creator who created it. But in any case, there are very, very great obligations upon me. I *do* have to save my soul. There is *not* just a holiday here on earth, nor is it all just a meaningless thing. There is some . . . some *accountability,* which is pretty tough to carry out."

Enter Mrs. Schumacher, an attractive woman in her early forties, pink-cheeked, cordial. Would we care for tea? Any word from Robert? Schumacher asks. He just arrived, Mrs. Schumacher says—he is their sixteen-year-old son just back from California. Schumacher is delighted with this news. Yes, we'll have tea. *I do not know how to deal with this man.*

"Well, granting that you don't know how we got here, you—your book does have some very specific ideas about what we should do about being here?"

"The publication of it changed my life. There was no desk work in it. I do not consider myself so much a writer as a worker with ideas." He is easing off now. Either the question, or the news of his son, or the fact that he has got a lot off his chest, quiets him down. "It is simply a collection of essays which arose from living situations, which we saw to be interconnected. It was a sort of Marxian insight in that Marx has taught us not to pay too much attention to what people say, but to study how they live. I don't hold to an economic interpretation of history—people may live rapaciously and still have beautiful theories—but to study how people *live* is a good thing. And the main element of living now is technology.

"Technology for the little people doesn't *exist* anymore." The subtitle to Schumacher's book, it is worth noting, is "Economics As If People Mattered." He goes on: "It has virtually disappeared except for the lowest level of technology, which is indestructible—primitive hand tools." In a curious and wholly unexpected way (a way he surely would reject if he were made aware of it) the man is circling back toward the earth-grubbing Leakey. "What you have now," Schumacher continues, "is a technology that is only for multinational corporations." What he is interested in, he has said elsewhere, is the technology between the hoe and the tractor.

"Twelve years ago, some of us set up an organization to fill this missing middle I speak of. In other words, to increase the technological *options.* Now I like to think there are a lot of answers to a lot of problems in this approach."

There is considerable evidence that in this respect, at least, Schumacher is right. In 1965, Schumacher put together out of the technology he deplores a group of like-minded experts whose purpose was to find an alternate to the "bigger is better" approach—a low-cost and labor-intensive technology in marked contrast to the heavy infusion of capital and equipment that replace men with machines. From a group of engineers, scientists, architects, doctors, building technicians, and other specialists, the Intermediate Technology Group was set up, first in London, and now throughout the world, to solve specific problems arising in underdeveloped nations. Such problems should order their own solutions, they believed. Since then, these experts have produced low-cost farm machinery and hospital equipment; a small-scale sugar refinery in Sri Lanka; ferro-cement boats in Sudan that the Sudanese can build themselves; miniturbines for the Pakistanis to adapt to small streams flowing out of the Himalayas. "All you need is simple materials," Schumacher once told a reporter. "The Taj Mahal wasn't built with Portland cement."

His own experience as a farm laborer in England during World War II had told him that solutions are often more obvious than they might seem. "We worked on a three-hundred-acre farm, and we used mainly animal-drawn equipment. We farmed very efficiently, we had very good yields, and the total equipment used cost about ten thousand dollars at present-day prices. Not a single piece of that equipment exists any more. To farm that same three hundred acres now you'd have to spend a hundred or two hundred thousand. To find ten thousand dollars is *one* thing—maybe you and I could do it—but to find two hundred thousand is quite a *different* thing. The circle of people admitted to the dinner table is narrowed thereby. So we said, intermediate technology is *disappearing* again, polarizing society into the rich and the *helpless.* And no matter *what* we do the gap becomes bigger and bigger, even in the U. S. The helpless people are *totally* helpless. They're kept afloat by welfare, and now the entire society is groaning under this ever-increasing load. They are not being rehabilitated, reintegrated. So we said we must find a way to use *present* knowledge to create a technology with more self-help for the people who do *not* have two hundred thousand dollars. In other words, to increase the technological *options.*"

One of the best-known innovations of Schumacher's group is a machine for making Zambian egg crates. In the West, the smallest device used for this purpose costs $390,000, with a much greater production capacity than Zambia required. Schumacher's group designed one to do only what was required, for $19,500.

A programmatic solution went with the hardware. If work was to be the answer (and for Schumacher it most certainly is):

—Work will have to be carried to the people wherever they might be, rather than remaining centered in the cities where there are now too many people and not enough jobs.
—Tools will have to be cheap so as to be affordable.
—Work will be simple enough for anyone to do.
—Production will arise from local materials adapted to local use.

Schumacher picks up from the table a copy of his new book, *A Guide for the Perplexed.* "My publisher sent me a copy by airmail express for twenty-five dollars. Twenty-five dollars! Then I received another copy—by airmail express. Why? I can understand the urgency of getting me the first copy, and I am grateful for this, but why must they send the next copy the same way? Why not sea-freight it at a reasonable cost? I already *have* the first copy. In large organizations you can't keep in touch with reality."

"Were your ideas intended only for the problems of the Third World?" I ask him.

"Yes. Exclusively for the Third World at first. But since the beginning of the fuel crisis in 1973, people say, 'Why do you waste your time on the Third World? What about *us?*' So that even in the United States now, Congress has set up the National Center for Appropriate Technology, which has nothing to *do* with the Third World. It tries to grapple with the problems of the polarization of American society. Forty million people are out of the mainstream and nobody knows how to reintegrate them. There are two polarizations—one is between rich and poor. If you're with it, okay, but if you're not, you'll never get back into it. And the other is geographical, between a few places where the action is—the cities—and all the rest where it is not."

Just now, it is the problems of the cities that seem most to concern him. "In America, there are now some thirty or forty million people trapped inside them. Many of them want to get out, but they have nothing to get *into.*" He slips again into his one-man dialogue, but less tendentiously.

" 'Where do you come from?' I say.

" 'From Bismarck, North Dakota.'

" 'Oh, you like it better in New York?'

" 'Oh, no, I *don't.* I'd like to get back with my children and my wife to Bismarck, North Dakota.'

" 'And I say, 'Why don't you *do* it, then?'

" 'I'd like to get out of Sydney, back into Tasmania—no jobs!' 'Out of Toronto into Saskatchewan. *No jobs!*' That's why Bismarck and Tasmania and Saskatchewan all say, 'The things we used to make ourselves are now coming in cellophane packages from Chicago, or Toronto, or Sydney!'

"I've coined a term—it's 'internal colonialism.' These regions are now *colonies* of the big cities!"

In Montana, the governor asked Schumacher what was to become of his state. Of 100 university graduates, eighty-five leave for jobs elsewhere. The governor told Schumacher Montana is now an underdeveloped country, like India.

"He says to me, 'We are looked upon by the rest of the country as a resource. They see there is a lot of coal here and say, 'let's rip it up.' So the governor of Montana has imposed a coal severance tax, telling the United States: 'At least we want to make money.'

"And the governor asked me, 'What should we do with this money? What do we do with it to develop Montana, with its small population of seven hundred thousand?'

"But Montana as a place *in itself* is worth something. If it were a separate country, people would not look upon it simply as a place out of which you rip *timber* and *wheat* and *coal*—they would respect it as a *culture,* help it to become a co*her*ent *thing!* The same is true of Oregon.

"Well, I could name you many other states. *Vermont!* 'We do *not* just want to be a playground and colony of New York City. We have a proud culture of our own. We want an interesting life here. What can we make of Vermont?' " In upstate New York, I reflect as Schumacher moves rhetorically across the nation, I live in a similarly colonial setting, rolling hills to the west of the Hudson, but a depressed county whose principal jobs are held by schoolteachers and prison guards. Schumacher is coming very close to *my* home.

"What do you tell them?" I asked.

"That all of life is a matter of structures, something modern liberalism has tried to refute. There is some *limitation! Look upon Vermont as a foreign country!* Find out, economically speaking, what goes out and what comes in. The first thing you find is that Vermont is still dairy land. So they produce a *lot* of milk. What happens to it? It is all exported in great tankers down to New York, there to be pasteurized, stabilized, homogenized, vitamin-reinforced, put into bottles—and re-exported to *Vermont!* So, not surprisingly, you have *no jobs* in Vermont!

"Why don't you do that yourself ? " Schumacher says to Vermont. "The first answer is that, 'Unless these things are done on a super, super scale, they're not worth doing.' Is this really so? Have you really looked at it? In every part of the world, there is a sufficient variety of jobs for people to have an interesting life."

"You believe this is possible *any*where?"

"Yes," he says evenly. "Being a European with a Swiss wife, Switzerland is all right. It's only five million people. Iceland is working. Norway, with two or three million. You don't have to *have* a hundred million."

"Is this your version of what some refer to as postindustrial civilization?"

"Yes. The competition now between capitalism and communism goes in the wrong direction. The fascination is with things rather than the quality of life. Not so long ago I was in East Germany, where they explained to me at great length that the Western system was inefficient compared to theirs. One of them said, 'The Western economies are like an express train moving at an ever-increasing speed straight into an abyss. But we shall overtake them!' " Nor is Schumacher necessarily unmindful of the irreplaceable nature of such modern utilities as power dams, television, and transatlantic airlines. "People think I want to take their technological toys away," he once remarked to another reporter. "I have no such intention. After a talk I made in Germany, as I was walking out an enraged couple followed behind me. The woman said, 'What stupid stuff this guy was talking. Do you know he arrived by airplane? How could he come by airplane if we only had intermediate technology?' Well, this was too much for me, and with the utmost politeness I turned around and said, 'Madam, in that case there would have been no *need* for me to come.' " He smiles tightly.

Like everyone else he was brought up to believe that in order to effect change, the "system" must change. But this in fact has never been so. The only change that is meaningful happens in the "real world," he says. "You have to make a viable future visible in the present. So that if you want to change agriculture away from chemicals and back to the organic, *then you only have to do it!*"

"You don't seek to attack the problem in the marketplace?"

"To do anything you have to be able to *survive* in the marketplace. Still there will be resistance. If the organic farmer is viable, people will say, 'Well, *he* gets away with it but of course it's not for everyone. There could never be *enough* organic material. Why change?'

"Well, we have a change when what we have been doing *doesn't work* any more. Why change from fossil fuels? Because they will be *running out!* And so the pressures mount. People get worried—fertilizer prices are going through the roof! Soon they will not be purchasable. *Then* they will remember there are these oddballs who get good yields without fertilizer. 'How the hell do they do it?' Suddenly the wind blows into *your* sail. Then change can be very rapid." It is a first conjunction with Hardin, I note, who had said the same with respect to antiabortion laws.

But where, I persist, does change start? At the government level?

"Yes and no," Schumacher says. "There must be some *people* who will start it. I believe you always have to aim at the ABCD combination."

"ABC . . . ?"

"*A* are administrators, government people. Try to have them with you in the group you form because they know the ropes. They can't do much, but they could stop everything. It's better to have them on your side.

"*B* is for business. You have to have business in it. The businessman produces everything you use, no matter how prejudiced intellectuals are against him. He is the man of viability.

"*C* is for the communicators, people of the word. They don't produce bread and butter and clothes, but they solve problems, and they have ideas. Research people, academics.

"*D* is for the democratic organizations of society. The troops, as it were. Women's organizations, unions, and so on—to put it all out to the masses. All this is irrespective of whatever political system exists within the country. Yes, and in every one of our centers—we have twenty-six about the world—we try to get this combination."

My God, I thought, this surprising, near-fanatic, simplistic, contradictory, inspired British-German with his ABC's really *believes* he can save the world!

Quarreling with his methods and his reasoning—those of a capitalist entrepreneur and a religious fundamentalist—a young American counterculture writer sees him nevertheless as having "the overarching virtue of being the man who articulated for our time, in at least initial form and with considerable wit, the philosophy I believe is the most important in the world today"—that is, that to live in some degree of compatibility with his environment, man must reshape his institutions and his technology so that they are cheaper, safer, simpler —smaller. He must rethink his place in the world. But such a troublesome prophet Schumacher must be to follow, and how much of importance he is ready to ignore!

"Have agencies of the United Nations—UNEP or UNESCO—availed themselves of your methods?" I asked him.

"They employ a lot of nice people of goodwill. They visit our London office, and they say, 'What you are doing is of the greatest possible interest.' But as organizations go, I find them totally impotent."

"Is this because of the lack of sovereign power?"

"Could be." He shrugged. "They are too big, too remote."

"What are your sources of funding?"

"Very little. In America, the Lilly Foundation. But we have developed ways to do things very cheaply. We scrape funds together according to the project. We are basically a *knowledge* organization"—a second echo of Hardin's notions.

"But getting back to the point," he says now, rather too eagerly, "I said in my chapter on education in *Small Is Beautiful* that we can do all sorts of things, but unless we get our *basis* right—I called it a metaphysical reconstruction—even the *positive* things will only add to the general confusion. Because they will not remain *positive!* I am talking about metaphysics!" With some dread, I realize he is climbing back into the pulpit; I begin to look for a way to head him off. "To return to where I started, if we plug this line—this *non*science parading as science that I call a *hoax*... that this is a *meaningless* universe with no higher intelligence behind it, a collocation of atoms and chance mutations sorted out by survival . . . if *this* is our basic conviction and we plug this and pump it into everyone and then on Sunday we talk about goodness, human rights . . . 'Well,' "—the antagonist is back—" 'An *accident* has no rights, that's *ridiculous!*' Then the metaphysical confusion will be so great that *everything will be debased.* Everything becomes a bandwagon because the only thing that remains is physical comfort, or excitement. And we will be *debased* into some *utilitarian* thing. People will say, 'They are setting up a Center for Appropriate Technology in Butte, Montana!'

" 'How much money is there in it?'

" 'Oh, a budget of three million a year.'

" 'Well, let's see if we can get in on the *action!*' "

It is perhaps wrong of me to think so, but I believe he is baiting me now. I have come too long a way, both in distance and the information I have acquired, and there is not much time left. Evidently sensing my resistance before I have had the chance to protest, he says politely, "Please, if I am talking too critically . . ."

"But what do you see as a counter to this thinking? You say metaphysics. What do you mean by metaphysics? Metaphysics as a guide to right and wrong? *What* right and wrong?"

"Yes, exactly. Yes." He has not expected a challenge and he is as embarrassed at having to answer it as I am at having to raise it. Neither of us seems to know which way to go next. "This whole problem in my view," he says, "should show the real humanity, which is *not* physics. The real concerns of people are mainly interpersonal relations. Not the *Big Bang,* not these doubts continuously cast on human beings. . . "

"But if people today live in a state of confusion and are committed to the false values suggested by science, what values should they live by? As defined by whom?"

Silence.

"Because if you suggest religion as the way, religion in the conventional sense, then of course you have the precedents of religious wars and their destructiveness, and the misinterpretations of religion. Some believe the very act of ravaging natural resources derives from religious injunctions—that God put the earth here for man to exploit. You have war, crime, and destruction committed in the name of various religions. Doesn't confusion exist there for you as well?"

Silence.

"I read from your book a great personal and spiritual awareness of the innate responsibility of one individual to another, and to all of society's modern problems. But, as a program for living, most of what I get from your book is a pragmatic directive for remedy of the problems of technology. As a program for right and wrong, I don't see that. And I don't understand what you mean when you say we should ignore the materialism that comes from science to listen to—what?"

"Oh, no," Schumacher says, "I don't say that. That would be Manichaeism . . ." He leafs through the book before him, his new book, *A Guide for the Perplexed,* as though to find an answer for both of us. He tries to talk and read at the same time. It is an awkward moment, and I don't feel very good about it.

"I'm fully aware of it," he says, fumbling the pages, "that after we have been belting along for three hundred years increasingly into materialism . . . and that we are in such a dilemma . . . but occasionally you see signs of recovery, of health. Health, beauty, and permanence."

"But how do you codify the sort of behavior you believe to be healthy?"

"Read it out of what is actually happening. Out of the great moral teachings of religions. Now if I may say so, to say that there have been religious wars, well, there *have* been wars, there *have* been crimes throughout the history of mankind, and if the main interest of people is in religion, then the wars and crimes have assumed that color."

Elsewhere I later read a summation by Arnold Toynbee that does for Schumacher's mysticism what he is not so well able here and now to do for himself:

> These religious founders (Jesus, Buddha, Lao Tse, St. Francis) disagreed with each other in their picture of what is the nature of the universe, the nature of spiritual life, the nature of ultimate spiritual reality. But they all agreed that the pursuit of material wealth is a wrong aim. We should aim only at the minimum wealth needed to maintain life; and our main aim should be spiritual. They all said with one voice that if we made material wealth our paramount aim, this would lead to disaster. They also spoke in favor of unselfishness and of love for other people as the key to happiness and to success in human affairs.

Schumacher says: "*Every*thing is *not* permissible. Yesterday in *The Times* I read about a group wanting in the name of progress to legalize sexual relations between adults and children. The members said that these are urges some people have, and so long as the children consent, what is wrong with it? 'As long as you want something, it's all right.' In other words there are no values anymore. Well, we cannot survive like that! 'Anything' is anything *permissible.* Where does *anybody* draw the line? I've had such arguments with physicists. They want to put plutonium into the world, and they say, 'It's perfectly all right. What's wrong with it?' Is there anything wrong with anything? I ask."

Grasping a straw to help us both out, I mention that such questions seem to be emerging among those in America known as "radical biologists."

"Exactly!" he says. "There is now a growing moral responsibility. So simply to say how, and where is it codified?—*that* is our task and we have to *find* such things. The help we have, traditionally speaking, is revelation and nature."

We will get no further on this question, and there are only a few minutes before I must leave to catch my train. I turn the conversation back to the more immediate problems of ecology, and the imponderables surrounding the dilemma of tropical forests.

"I have just come back from western Australia," he says. "They have some of the most remarkable, incredible forests. Virgin forests. Various kinds of eucalyptus—the only comparable trees are sequoias in the United States. Five, six, seven hundred years old, two hundred feet high! Unbelievable trees. Most of them have been sold to the

Japanese to turn into chips. Not even for chipboard"—he laughs—"for *pulp!* For newsprint and wrapping paper. Somebody has been conned.

"Chips are a good idea—for the offal. If you saw out a lot of timber, don't burn the offal, use it for chips. But in western Australia there are big chippers that accept only the straightest and cleanest timber. So that 60 percent of the wood isn't even turned into chips. It's pushed together by bulldozers and burned!

"Here again, what one is up against is the question, 'Why not? Business is business.' The Japanese say, 'We need the stuff,' and the Australians say, 'Why should we worry about posterity? What has posterity ever done for us? *Don't* come to us about gene pools. That has always happened. Look at the list of extinct species. Who misses the dinosaur?' " He laughs. " 'So who will miss the Australian mahogany?' It really is *vandalism!* 'What vandalism? Don't introduce such unscientific terms. This is an *economics* proposition!'

"You have to appeal to the *whole* human being," he says. "The logical mind is only a small, little executive part of the whole human being."

"The little executive part?"

"Yes. After you know what you want to do, which is a supra- or metalogical question, then *how* do you do it? Your point of departure has been chosen, and your aim has been chosen. That's (a) and (b). Well, how do you connect (a) and (b)? That's what your logical faculties can do. But in the *choice* of (a) and the *choice* of (b), the *whole* human person comes in, with what they *used* to call the intellect. Which was *everything.*"

In his backyard, Schumacher has recently set out eight young trees. He walks me out to have a look at them and remarks that the Buddha taught that every man should be responsible for planting eight trees within his lifetime. "An American told me that only 10 percent of the earth is fit for agriculture," he says. "But there is plenty of room left over for trees—food trees. With our intelligence about genetics today, there's no reason we should not be growing trees. They are no trouble." He doesn't go on about this, but I later read in his book this imperative as a way to make a simple beginning with India's environmental problems. "Every educated man's first responsibility is to those uneducated who make his privilege possible," he writes. "What sort of an education is this if it prevents us from thinking of things ready to be done immediately? What makes us think we need electricity, cement, and steel before we can do anything at all?"

"Never ask a man to act against his own self-interest," Garrett Hardin had said. And the answer, if there should be an answer, lay for individuals as well as nations, in self-reliance. Not self-*sufficiency,* Hardin insisted, self-reliance. He had said, too, that he agreed with this man Schumacher not in fundamentals but in secondary aspects of his ideas. There is no conflict that I can see thus far (programmatically, at least) except that Schumacher has not once referred to the single threat to human survival which Hardin saw above all others—the crushing weight and the continuing swell of human population. Nor do I recall now having seen very much about it in his book. When I ask him his opinion on the question, he says, "Yes, uh . . ." and after a long pause, "the Buddha said, 'Work out your salvation with diligence, and don't worry about questions and problems that in fact you can't do anything about."

He walks me back toward the house—I have left my briefcase in his living room. "I know I arouse terrible criticism when I say this, but I am not a gynecologist. This is not my *métier.* The problem of world population does not exist."

"That isn't exactly what I've been led to understand," I said.

"There is no such thing as a *world* population. No. It is *not* an intercommunicating system. People who say 'six billion people! How awful, how dreadful!' They just don't know *what* they are talking about. They are not talking about reality."

"What is the reality of the population problem?"

"There are specific problems in Mauritius, in Japan, in Bangladesh. But you see, even in Bangladesh there is enough land to feed the population, there is no doubt about *that.* There are not the *resources* to keep them going. If you go to northern California, you'll find a French intensive biodynamic horticulture and agriculture system that can feed people in climatic conditions which are vastly inferior to those of *Bangladesh.* On twenty-five hundred square feet a person! What must be studied are the methods of working with nature. From the point of view of feeding the world, American agriculture is not efficient. It is vastly wasteful. Twelve thousand square feet to feed one person, even in California. In India thirty-two thousand square feet are required. But it is *possible* on *twenty-five hundred square feet.* Talk of a population problem is a copout. You just talk, you don't *do* anything."

"What about Tanzania, with its population of fourteen million, where cultural traditions interfere with the implementation of agricultural procedures? Cattle are valued by many more as property than as a food source. The cattle overgraze the grasslands. People reproduce in great numbers to assure their survival. How can you simply say, let these people see the problem and then they'll solve it?"

"I'm not saying that. What I am saying is that there are *specific* locations in the world where there are thousands of problems. I am saying there is no *world* population problem. You are adding up things you are not entitled to add up. In Tanzania there are many *specific* problems. Overgrazing is a very real problem. If in Tanzania people are putting into the world more children than can be supported, well, that is for the Tanzanians to say. It is a problem that is not going to solve itself, but *you and I can't solve it!*"

"Well, we have tried—by giving food. Don't such problems move toward what you call metaphysics? If we could make a difference, wouldn't we have a higher responsibility to try to do so?"

"The Point Four deliveries? To Tanzania? I don't know . . ."

"We delivered food to the sub-Sahel in the famine of 1974."

"Ah, a famine. Well, there you are talking about a disaster."

"But aren't you talking about *incipient* famine with the problems of nations like Tanzania?"

"This is such a big subject that I hesitate to deal with it swiftly for fear of creating a misunderstanding. There are disaster situations which can befall anybody. Then of course it is wonderful if we have air transport to mitigate them. There *are* very hard problems in Japan, Bangladesh, and India. Long-term problems like excessive population growth. I have just come from Indonesia. Java and Bali are very densely populated. But *not* that there's not enough room. There's an enormous amount of space. But there is this polarization tendency of mankind, which *I* think has to do with technology—the tendency for people all to go to *one place!*

"In the United States you have 92 percent of the population with all their appurtenances on *2* percent of the surface area—so the other 98 percent is virtually empty, and *not* because it is uninhabitable. If you would put the whole world population into the United States, you would have a population density roughly the same as England has now. So, if there are problems of overgrazing, let's attend to them. But there is also the problem that societies must somehow take responsibility for themselves. If you go with a blank check and say it is the duty of the United States to feed Tanzania or Barbados, how do you ever get to the point of waking them up? Of convincing them *they* must be responsible!

"You introduce a sort of worldwide *welfare* situation," Schumacher goes on. "I am in favor of welfare, but I am also bound to recognize that if welfare becomes a way of life, it destroys human responsibility. You have in your country people who are third-generation welfare recipients. And they are virtually irretrievable. They have never known anything other than the welfare check. It is a tricky thing. That's why I worry about talking too swiftly about it and we leave each other full of misunderstanding. The thing to do is to select subjects where you are competent and *entitled* to do something."

"You seem to be saying everything Garrett Hardin says but in another way. For him it is the carrying capacity of a country which is the measure, and from this he derives his lifeboat ethic . . ."

"Yes," Schumacher says, frowning, "but I think the way this has come across is quite deplorable. I'll have nothing to do with it. His way is that the problem is quantified, the real *quality* of it taken out. As if a nation *were* just a lifeboat!" He hisses: "It *isn't!* Tanzania is potentially a vastly fertile country. It has never been looked at from this point of view. The Europeans have come in and said, in this arid climate you can grow sisal, so let's grow sisal in Tanzania. We do *not* understand African agriculture. We understand agriculture in temperate climates.

"You simply cannot treat such things as a problem of technology —of birth control and abortion. One really doesn't know what one is doing to these societies, because they are extremely *subtle* problems. I've seen how it's likely to misfire. Mrs. Gandhi fell into this trap. *Particularly* when it comes from someone like ourselves who are grabbing fifty times the resources these people need to live.

"Very often, anything you try to do is totally counterproductive because it arouses bad feelings, as we saw at the World Population Conference in Bucharest. 'Not only are they stripping the world, but now they are going to begrudge us our children.' "He laughs." 'Even *that!*' So it must come from *inside,* from people realizing what must be done." As though on cue, a child's cry comes from a back room. "Mind you," he says, "I'm not saying this because I've got eight children myself. I'm very conscious of this. *They are my responsibility!* And my God! It's a big burden."

I am not the person to debate this man. Hardin should be here to speak for himself, or any number of the others, all of whom would vigorously resist the discrete analysis he insists on. And yet, if Schumacher should happen to be right—if fourteen million Tanzanians should insist on solving their own problems (as Julius Nyerere *has* insisted all along)—what then? Is the world still so large a place that each of its problems could yield to a local solution? A slim hope, indeed, but the two men fundamentally opposed through the authority they cite—Hardin out of Darwin; Schumacher out of Buddha, Muhammad, and Jesus—*were* saying the same thing. Not self-sufficiency, as Hardin argued, self-reliance. In *Small Is Beautiful,* Schumacher says: "Give a man a fish, as the saying goes, and you are helping him a little bit for a very short while; teach him the art of fishing and he can help himself all his life." Give information, Schumacher argues; and so does Hardin.

I have collected my things. Mrs. Schumacher is to drive me to the station, and her husband follows us to the car, still preaching, to add a final word. "The whole attitude of those who want to do something assumes there is some sort of steering mechanism—if only we would realize that endless exponential growth leads us to disaster. Which of course everybody *does* realize! Only fools don't realize that. It isn't that kind of a *problem.* It is *not* a world problem. It's a specific problem. There has to be a far more decentralized responsibility. If you can help people to raise their self-consciousness, their appreciation of reality, help them understand they are *not* simply victims of big systems but they also have responsibilities . . . *a certain sovereignty in themselves.*" (He pauses a long time after this; Mrs. Schumacher has the motor running.) *"That they can arrange their lives . . ."*

We arrived at the station with only moments to spare. Ten days later, in New York City, I would read of Schumacher's death. Traveling through Switzerland by train, he was stricken by a heart attack, and he died instantly. His work would continue, of course, through the agencies he had founded. But here and now, I hurried aboard my train. At Victoria Station I would have to see if I could find some way, given the slowdown of air controllers, to get out of London and across the Channel. If I could make it to Brussels by tomorrow morning, I could probably get a flight out to Moscow by late afternoon and still arrive in time to see Oparin on Saturday. . . .

A NOTE UPON RETURNING

Vihtelic lay on the side of the road in the warm sun. Soon one of the trucks came, and when it drew closer, close enough for the driver to see him lying there, it stopped.

"What's the matter?" the driver said.

"My foot is hurt and I have to see a doctor," Vihtelic said.

"What happened to you?"

"My car went over the side."

"How long were you down there?"

"Two weeks."

The driver crossed to the other side of the road and looked down. "I don't see any car down there," he said.

Vihtelic got up, and with the branch he had fashioned into a crutch to help him in the climb out of the ravine, he joined the driver. Together they peered at the foliage below. The station wagon was hidden beneath the fir trees. They moved up the road to the point Vihtelic had aimed for with his racket mirror, the opening in the tree line where he had watched cars and trucks pass for the last fifteen days. Even from here they were unable to see the station wagon below.

On the way to the Ranger station the driver offered him a sandwich from his lunch pail. Vihtelic ate it slowly. At the station he went to the bathroom and threw his underclothes into the wastebasket. He took an orange soda from the vending machine and then he placed a call to his mother in Whitehall. There was no answer.

An ambulance arrived from White Salmon. A young intern administered a shot of morphine and said he guessed Vihtelic might not lose his foot after all, but Vihtelic knew he was wrong, that his foot would have to come off. They drove him to the small hospital at White Salmon, arriving at two-thirty. He was given another shot of morphine and a series of tests and X-rays and another shot of morphine. Then they put him in another ambulance and drove him sixty miles to a larger hospital in Portland where he was moved into a semiprivate room with a man named Gordie. Gordie's wife had brought him a television set. The two of them watched the Monday night football game, and then they watched the late show, and then the movie that came on after that.

Gordie went to sleep in the middle of it but Vihtelic was determined to stay awake. He was afraid that if he didn't he would wake up back in the station wagon.

Finally, at four in the morning of the seventeenth day, having fought against it as long as he could, John Vihtelic went to sleep.

BIBLIOGRAPHY

Adler, Irving. *How Life Began.* rev. ed. New York: The John Day Company, 1977.

Augusta, Joseph. *Prehistoric Animals.* London: Spring Books, 1957.

Bateson, P. P., and Hinde, R. A. *Growing Points in Ethology.* New York: Cambridge University Press, 1976.

Blueprint for Survival, the editors of *The Ecologist.* Boston: Houghton Mifflin Co., 1972.

Bronowski, J. *The Ascent of Man.* Boston: Little, Brown & Company, 1973.

Brown, Leslie. *Africa: A Natural History.* New York: Random House, Inc., 1965.

Brown, Lester R., McGrath, Patricia L., and Stokes, Bruce. *Twenty-two Dimensions of the Population Problem.* Worldwatch Paper 5, March, 1976.

Carr, Archie. *So Excellent a Fishe.* Garden City: The Natural History Press, 1967.

Cotton, Steve, ed. *Earth Day—The Beginning.* New York: Arno Press, Inc., 1970.

Coulter, Merle Crowe, revised by Howard J. Dittmer. *The Story of the Plant Kingdom.* Chicago: University of Chicago Press, 1959.

Crittenden, Ann. "Economist Here Thinks Small." *New York Times,* Oct. 26, 1975.

Cronquist, Arthur. "Adapt or Die!" *Bulletin du Jardin Botanique National de Belgique* 41 (1971): 135–144.

Curtis, Helena. *Biology* 3rd. ed. New York: Worth Publishers, Inc., 1979.

Dasmann, Raymond F. *The Conservation Alternative.* New York: John Wiley & Sons, Inc., 1975.

Dawkins, Richard. *The Selfish Gene.* New York and Oxford: Oxford University Press, 1976.

Dickerson, Richard E. "Chemical Evolution and the Origin of Life." *Scientific American* 239 (1980): 70–86.

Dittmer, Howard. *Phylogeny and Form in Plant Kingdom.* Princeton: Van Nostrand Reinhold Company, 1964.

Dubos, Rene. "Symbiosis between the Earth and Humankind." *Science* 193 (1976): 459–462.

Dunbar, Carl O., and Waage, Karl M. *Historical Geology.* 3rd ed. New York: John Wiley & Sons, Inc., 1969.

Eccles, John C. "The Physiology of Imagination." *Scientific American* 199 (1958): 135–149.

Ehrenfeld, David W. *Biological Conservation.* New York: Holt, Rinehart and Winston, Inc., 1970.

Elgin, Duane, and Mitchell, Arnold. "Voluntary Simplicity." *The CoEvolution Quarterly,* Summer 1977.

Evans, Howard Ensign. *Life on a Little-Known Planet.* New York: E. P. Dutton Publishing Co., Inc., 1968.

Ferris, T. "Crucibles of the Cosmos." *New York Times Magazine,* 14 January 1979, p. 29.

———. *The Red Limit.* New York: William Morrow and Company, 1977.

Field, George, Verschuur, G., and Ponnamperuma, C. *Cosmic Evolution.* Boston: Houghton Mifflin Company, 1978.

Fox, Sidney W., and Dose, Klause. *Molecular Evolution and the Origin of Life.* San Francisco: W. H. Freeman and Co., 1972.

Frankel, O. H. "Genetic Resources." *Annals of the New York Academy of Sciences* 287 (1977): 332–334.

———. "The Time Scale of Concern." In *Conservation of Threatened Plants,* edited by J. B. Simmons, R. I. Beyer, P. E. Brandham, G. Lucas and V. T. H. Parry. New York: Plenum Publishing Corp., 1976.

———. "Genetic Conservation in Perspective." In *Genetic Resources of*

Plants, edited by O. H. Frankel and E. Bennett. Oxford: Blackwell Scientific Publications, Ltd., 1970.

Gamow, George. *Biography of the Earth.* rev. ed. New York: The Viking Press, Inc., 1959.

Gardner, Martin. "Bang!" *New York Review of Books,* 12 May 1977, p. 29.

______. "The Holes in Black Holes." *New York Review of Books.* 29 September 1977, p. 22.

Gould, Stephen Jay. "An Unsung Single-celled Hero." *Natural History,* November 1974, p. 33.

______. *Ever Since Darwin.* New York: W. W. Norton & Co., 1977.

______. "Mankind Stood up First and Got Smart Later." *New York Times,* 22 April 1979.

______. "Our Greatest Evolutionary Step." *Natural History,* June 1979, p. 40.

Gribbon, Dr. John. "Cosmic Luck." *New Scientist,* 26 October 1978, p. 40.

Handler, Bruce. "The Politics of Water." *Saturday Review,* 14 May 1977, pp. 15–19.

Handler, Phillip. "On the State of Man." *BioScience* 25 (1975): 425.

Hardin, Garrett. "Are Technological Fixes Enough?" In *Global Chemical Cycles and Their Alterations by Man,* edited by Werner Stumm. Berlin: Dahlem Konferenzen, 1977, p. 335.

______. "Carrying Capacity As an Ethical Concept." *Soundings: An Interdisciplinary Journal,* Spring 1976.

______. *Exploring New Ethics for Survival.* New York: The Viking Press, Inc., 1972.

______. "Living on a Lifeboat." *Bioscience* 24 (1974): 561–568.

______. "Living with the Faustian Bargain." *Bulletin of the Atomic Scientists* 32 (1976): 26–29.

______. "Pejorism: The Middle Way." *The North American Review* 261 (1976).

______. "The Fallibility Factor." *Sceptic,* no. 14 (1976).

______. "The Rational Foundation of Conservation." *The North American Review* 259 (1974): 14–17.

Harlan, Jack R. "The Plants and Animals That Nourish Man." *Scientific American* 235 (1976): 89–97.

Harris, Marvin. *Cannibals and Kings.* New York: Random House, Inc. 1977.

______. "How Green the Revolution," *Natural History* June–July 1972; p.28.

Hayes, Harold T. P. *The Last Place on Earth.* Briarcliff Manor: Stein and Day Publishers, 1977.

———. "The Pursuit of Reason." *New York Times Magazine,* 12 June 1977, p. 21.

Heady, Earl O. "The Agriculture of the U. S." *Scientific American* 235 (1976): 30–39.

Hinde, Robert A. "The Nature of Aggression." *New Society* (London), 22 March 1967; pp. 302–304.

Hoagland, Hudson. "Brain Evolution and the Biology of Belief." *Bulletin of the Atomic Scientists* 33 (1977): 41–44.

Holloway, Ralph. Personal interview. October 1978.

Howell, F. Clark. *Early Man.* New York: Time-Life Books Inc., 1968.

Hutchinson, G. E. "Homage to Santa Rosalia, or Why Are There So Many Kinds of Animals?" *The American Naturalist* May–June 1959, pp. 145–157.

Jennings, Peter R. "The Amplification of Agricultural Development." *Scientific American* 235 (1976): 181–194.

Langer, Susanne K. *Mind: An Essay on Human Feeling.* Baltimore: The Johns Hopkins University Press, 1967.

Lauwerys, Joseph A. *Man's Impact on Nature.* Garden City: The Natural History Press, 1970.

Leakey, Mary D. "Footprints in the Ashes of Time." *National Geographic,* 155 (1979): 446–457.

Leakey, Richard E., and Lewin, Roger. *Origins.* New York: E. P. Dutton Publishing Co., Inc., 1977.

Leeper, E. M. "A Chat with Garrett Hardin: Master of Metaphor Expands 'Commons' Thesis." *BioScience* 26 (1976): 785–787.

Lorenz, K. *On Aggression.* New York: Harcourt Brace Jovanovich, Inc. 1966.

Lovins, Amory B. "Long-term Constraints on Human Activity." *Environmental Conservation* 3 (1976).

Luria, S. E. *Life—The Unfinished Experiment.* New York: Charles Scribner's Sons, 1973.

MacNamara, Robert S. "An Address on the Population Problem." Paper delivered at Massachusetts Institute of Technology, April 28, 1977.

Man and the Biosphere, Number 12, Morges: UNESCO, 25–27, September 1973.

May, Robert M. "The Evolution of Ecological Systems." *Scientific American* 239 (1978): 141–159.

Mayr, Ernst. "Evolution." *Scientific American* 239 (1978): 47–55.

McBride, Stewart Dill. "When E. F. Schumacher Talks. . . ." *The Christian Science Monitor,* 27 June 1977, pp. 14–15.

McRobie, George. "Intermediate Technology in Action." *Intermediate Technology Development Group, Ltd.,* Autumn 1976.

Medawar, P. B., and Medawar, J. S. *The Life Science.* New York: Harper and Row, Publishers, Inc., 1977.

Mesarovic, M., and Pestel, E. *Mankind at the Turning Point: The Second Report to the Club of Rome.* New York: E. P. Dutton Publishing Co., 1974.

Monod, Jacques. *Chance and Necessity.* New York: Alfred A. Knopf, Inc., 1971.

Montague, Ashley. *The Direction of Human Development.* New York: Hawthorn Books, Inc. 1970.

Moore, John A., et. al. *Biological Science.* 2nd ed. New York: Harcourt, Brace & World, 1968.

Moorehead, Alan. *Darwin and the Beagle.* New York: Harper and Row, Publishers, Inc. 1969.

Moss, Cynthia. *Portraits in the Wild.* Boston: Houghton Mifflin Company, 1975.

Myers, Norman. "An Expanded Approach to the Problem of Disappearing Species." *Science* 193 (1976): 198–201.

National Research Council. *Genetic Vulnerability of Major Crops.* Washington, D.C.: National Academy of Sciences, 1972.

Odum, Eugene P. "The Emergence of Ecology as a New Integrative Discipline." *Science* 195 (1977): 1289–1293.

Oparin, Aleksandr Ivanovich. *The Origin of Life.* New York: Dover Publications, Inc., 1953.

Passmore, John. *Man's Responsibility for Nature.* London: Duckworth, 1974.

Pfeiffer, John E. *The Emergence of Man.* New York: Harper and Row, Publishers, Inc., 1969.

Ponnamperuma, C. *The Origins of Life.* New York: E. P. Dutton Publishing Co., 1972.

Programme on Man and the Biosphere (MAB), *Conservation of Natural Areas and of the Genetic Material They Contain.* Paris: UNESCO, 1973.

Rees, Martin. "Galactic Nuclei and Quasars: Supermassive Black Holes?" *New Scientist,* 19 October 1978, p. 188.

——— and Silk, Joseph. "The Origin of Galaxies." *Scientific American,* 222 (1970): 26.

Rensberger, Boyce. "Rival Anthropologists Divide on 'Pre-Human' Find." *New York Times,* 18 February 1978.

Revelle, Roger. "The Resources Available for Agriculture." *Scientific American* 235 (1976): 105–178.

Rhodes, F. H. T. *The Evolution of Life*. 2nd ed. Baltimore and Middlesex: Penguin Books, 1976.

Romer, Alfred Sherwood. *The Vertebrate Story* (fifth impression). Chicago: The University of Chicago Press, 1967.

Roup, David M., and Stanley, Steven M. *Principles of Paleontology.* San Francisco: W. H. Freeman and Company, 1971.

Schopf, J. William. "The Evolution of the Earliest Cells." *Scientific American* 239 (1978): 111–138.

Schumacher, E. F. *Small Is Beautiful.* New York: Harper and Row, Publishers, Inc., 1973.

Simons, Elwyn L. "Ramapithecus." *Scientific American* 236 (1977): 28–35.

Simpson, George Gaylord. *Life of the Past.* New Haven: Yale University Press, 1953.

______. "The Origin of Mammals." In *Encyclopedia of Evolution,* edited by Bernhard Grzimek. New York: Van Nostrand Reinhold, 1976.

Smil, Vaclav. "Intermediate Energy Technology in China." *Bulletin of the Atomic Scientists* 33 (1977): 25–31.

Stirton, Ruben Arthur. *Time, Life, and Man.* New York: John Wiley & Sons, Inc., 1959.

Strong, Maurice F. "Baer-Huxley Memorial Lecture." Paper delivered at Second International Conference of Environmental Future, Reykjavik, Iceland, June 8, 1977.

Stunkel, Kenneth R. "The Technological Solution." *Bulletin of the Atomic Scientists* 29 (1973): 42–44.

Sullivan, Walter. "Scientists Hoard Old Grain Seeds." *New York Times,* 4 October 1970.

Taylor, Gordon Rattray. *Science of Life.* New York: McGraw-Hill, Inc. 1963.

Thenius, Eric. "The Tertiary Period—The Age of Mammals." In *Encyclopedia of Evolution,* edited by Bernhard Grzimek. New York: Van Nostrand Reinhold, 1976.

Thomas, William L., Jr. ed. *Man's Role in Changing the Face of the Earth*. Wenner-Gren Foundation for Anthropological Research, and National Science Foundation. Chicago: The University of Chicago Press, 1956.

Thornton, Mary. "Food Exports Threatened by Loss of Farmland in U.S." *The Washington Star,* 26 November 1976.

Thorpe, W. H. *Animal Nature and Human Nature.* Garden City: Anchor Press, 1974.

Valentine, James W. "The Evolution of Multicellular Plants and Animals." *Scientific American* 239 (1978): 141–159.

Wade, Nicholas. "E. F. Schumacher: Cutting Technology Down to Size." *Science* 189 (1975): 199–201.

______. "Green Revolution (I): A Just Technology, Often Unjust in Use." *Science* 186 (1974): 1093–1096.

______. "Green Revolution (II): Problems of Adapting a Western Technology." *Science* 186 (1974): 1186–1188.

Wagner, Richard H. *Environment and Man.* New York: W. W. Norton & Company, Inc., 1971.

Walker, Alan, and Leakey, Richard E. F. "The Hominids of East Turkana." *Scientific American* 239 (1978): 54–66.

Washburn, Sherwood J. "The Evolution of Man." *Scientific American* 239 (1978): 194–207.

Watt, Kenneth E.F. "Man's Efficient Rush Toward Deadly Dullness." *Natural History*, February 1972, p. 74.

______. "The End of an Energy Orgy." *Natural History,* February, 1974, pp. 16–22.

______. "The Long Arm of Biological Law, or How Charles Darwin and His Lot Will Inherit the Earth." *Natural History,* April 1971, pp. 14–19.

Webster, Bayard. "Wildlife Preservation Theory Challenged." *New York Times,* 31 May 1979, p. B13.

Weinberg, Steven. *The First Three Minutes: A Modern View of the Origin of the Universe.* New York: Basic Books, 1977.

Wendt, Herbert. "In the Deep Freeze of Nature." In *Encyclopedia of Evolution,* edited by Bernhard Grzimek. New York: Van Nostrand Reinhold, 1976.

White, Edmund, and Brown, Dale. *The First Men.* New York: Time-Life Books, 1973.

Wilkes, Garrison. "The World's Crop Plant Germplasm: An Endangered Resource." *Bulletin of the Atomic Scientists* 33 (1977): 8–16.

Wortman, Sterling. "Food and Agriculture." *Scientific American* 235 (1976): 30–39.

INDEX